Monitor

D0765100

Educational

Committee on Developing Indicators of Educational Equity

Christopher Edley, Jr., Judith Koenig, Natalie Nielsen,
and Constance Citro, *Editors*

Board on Testing and Assessment
and
Committee on National Statistics

Division of Behavioral and Social Sciences and Education

A Consensus Study Report of

The National Academies of
SCIENCES · ENGINEERING · MEDICINE

THE NATIONAL ACADEMIES PRESS
Washington, DC
www.nap.edu

THE NATIONAL ACADEMIES PRESS   500 Fifth Street, NW   Washington, DC 20001

This activity was supported by the American Educational Research Association (unnumbered), the Atlantic Philanthropies (Award No. 23223), the Ford Foundation (Award No. 0145-1710), the Spencer Foundation (Award No. 201500103), the U.S. Department of Education (Award No. R305U150002), the William T. Grant Foundation (Award No. 184262), and the W.K. Kellogg Foundation (Award No. P3033235). Any opinions, findings, conclusions, or recommendations expressed in this publication do not necessarily reflect the views of any organization or agency that provided support for the project.

International Standard Book Number-13: 978-0-309-49016-0
International Standard Book Number-10: 0-309-49016-2
Digital Object Identifier: https://doi.org/10.17226/25389
Library of Congress Control Number: 2019947474

Additional copies of this publication are available for sale from the National Academies Press, 500 Fifth Street, NW, Keck 360, Washington, DC 20001; (800) 624-6242 or (202) 334-3313; http://www.nap.edu.

Printed in the United States of America

Suggested citation: National Academies of Sciences, Engineering, and Medicine. (2019). *Monitoring Educational Equity.* Washington, DC: The National Academies Press. https://doi.org/10.17226/25389.

The National Academies of
SCIENCES · ENGINEERING · MEDICINE

The **National Academy of Sciences** was established in 1863 by an Act of Congress, signed by President Lincoln, as a private, nongovernmental institution to advise the nation on issues related to science and technology. Members are elected by their peers for outstanding contributions to research. Dr. Marcia McNutt is president.

The **National Academy of Engineering** was established in 1964 under the charter of the National Academy of Sciences to bring the practices of engineering to advising the nation. Members are elected by their peers for extraordinary contributions to engineering. Dr. C. D. Mote, Jr., is president.

The **National Academy of Medicine** (formerly the Institute of Medicine) was established in 1970 under the charter of the National Academy of Sciences to advise the nation on medical and health issues. Members are elected by their peers for distinguished contributions to medicine and health. Dr. Victor J. Dzau is president.

The three Academies work together as the **National Academies of Sciences, Engineering, and Medicine** to provide independent, objective analysis and advice to the nation and conduct other activities to solve complex problems and inform public policy decisions. The National Academies also encourage education and research, recognize outstanding contributions to knowledge, and increase public understanding in matters of science, engineering, and medicine.

Learn more about the National Academies of Sciences, Engineering, and Medicine at **www.nationalacademies.org**.

The National Academies of
SCIENCES · ENGINEERING · MEDICINE

Consensus Study Reports published by the National Academies of Sciences, Engineering, and Medicine document the evidence-based consensus on the study's statement of task by an authoring committee of experts. Reports typically include findings, conclusions, and recommendations based on information gathered by the committee and the committee's deliberations. Each report has been subjected to a rigorous and independent peer-review process and it represents the position of the National Academies on the statement of task.

Proceedings published by the National Academies of Sciences, Engineering, and Medicine chronicle the presentations and discussions at a workshop, symposium, or other event convened by the National Academies. The statements and opinions contained in proceedings are those of the participants and are not endorsed by other participants, the planning committee, or the National Academies.

For information about other products and activities of the National Academies, please visit www.nationalacademies.org/about/whatwedo.

COMMITTEE ON DEVELOPING
INDICATORS OF EDUCATIONAL EQUITY

CHRISTOPHER EDLEY, JR. (*Chair*), School of Law, University of
California, Berkeley
ELAINE ALLENSWORTH, UChicago Consortium on School Research,
University of Chicago
ALBERTO CARVALHO, Miami-Dade County Public Schools, Miami, FL
STELLA FLORES, Steinhardt School of Culture, Education, and Human
Development, New York University
NANCY GONZALES, College of Liberal Arts and Statistics, Arizona
State University
LAURA HAMILTON, RAND Corporation, Pittsburgh, PA
JAMES KEMPLE, The Research Alliance for New York City Schools and
Steinhardt School of Culture, Education, and Human Development,
New York University
SHARON LEWIS, Council of the Great City Schools, Washington, DC
(retired)
MICHAEL J. MACKENZIE, Centre for Research on Children and
Families, McGill University, Montreal, Canada
C. KENT MCGUIRE, William & Flora Hewlett Foundation, Menlo
Park, CA
SARA MCLANAHAN, Department of Sociology, Princeton University
MEREDITH PHILLIPS, Departments of Public Policy and Sociology,
Luskin School of Public Affairs, University of California, Los Angeles
MORGAN POLIKOFF, Rossier School of Education, University of
Southern California
SEAN F. REARDON, Graduate School of Education, Stanford University
KAROLYN TYSON, Department of Sociology, University of North
Carolina at Chapel Hill

JUDITH KOENIG, *Study Director*
NATALIE NIELSEN, *Senior Program Officer*
CONSTANCE CITRO, *Senior Scholar*
KELLY ARRINGTON, *Senior Program Assistant*

COMMITTEE ON NATIONAL STATISTICS

Preface

The challenge of monitoring disparities in educational achievement and opportunities shares some characteristics with other complex regulatory problems. For example, when Congress adopted the Clean Air Act (1970) nearly 50 years ago, it emphasized the importance of public health but provided no clear line for distinguishing clean air from dirty air. Most fundamentally, regulating pollution has required choices about what indicates that air is "polluted" for regulatory purposes, how to measure and monitor those indicators, and when the measured level of an indicator should trigger enforcement or other intervention. The statute provided few answers, or even a definitive list of "pollutants" to be regulated. Nor were there definitive answers in the Constitution, economics, the biological sciences, or epidemiology. Instead, definitions and decisions have been a continuous enterprise involving interpretations of vague statutory language, promulgation of hundreds of federal and state regulations, enforcement experience, research in multiple disciplines, and the turbulence of politics.

So it is with "regulating" educational equity and inequity—distinguishing between the good and the problematic in a system that powerfully shapes socioeconomic opportunity, outcomes, and mobility. For a century following the Civil War, the issue was largely a matter of antidiscrimination litigation, based on the U.S. Constitution. This is what can be thought of as *constitutional equality*. Beginning with the Elementary and Secondary Education Act in 1965, however, Congress and the executive branch built a broader, complementary framework for an evolving social policy construct of *regulatory equity*. Given the extensive disparities that still exist in the nation's educational system, what can policy makers do to better support goals for a just and prosperous society? What evidence can best inform their decisions? Specifically, if an educational equity construct is to have practical use, policy makers must choose *indicators and measures.*

This report provides the architecture for a system to help policy makers address questions of educational equity. It not only lays out a system of indicators of educational equity, but also describes some of the follow-on work needed to advance such a system through public consensus, engineering, construction, and continuous maintenance. The closest analogy in the education realm is probably the history the National Assessment of Educational Progress (NAEP), whose planning began in the early 1960s, was first fielded in 1969, has become a trusted measure of the knowledge of U.S. students, and continues to evolve.

This report would not have been possible without the contributions of many people. On behalf of the committee, I extend our deepest appreciation to the sponsors of this work: the American Educational Research Association, the Atlantic Philanthropies, the Ford Foundation, the Spencer Foundation, the U.S. Department of Education, the William T. Grant Foundation, and the W.K. Kellogg Foundation. Without their support, this study would not have come to fruition.

We also thank the experts who volunteered their time to share their knowledge with us: Ilene Berman, the Annie E. Casey Foundation; Betsy Brand, American Youth Policy Forum; Catherine Lhamon, Chair, U.S. Commission on Civil Rights; Nat Mulkas, American Enterprise Institute; David Murphey, Child Trends; Amber Northern, Fordham Institute; Jennifer Park, Federal Interagency Forum on Child and Family Statistics; Douglas Ready, Columbia University Teachers College; Natasha Ushomirsky, Education Trust; and Stephanie Wood-Garnett, Alliance for Excellence in Education.

The committee commissioned a set of experts to author literature reviews to help us identify indicators. We thank those writers for their invaluable input: David Campbell, University of Notre Dame; Jennifer Jennings, Princeton University; Katherine Magnuson, University of Wisconsin-Madison; Nicholas Mark, New York University; Jenny Nagaoka, University of Chicago Consortium on School Research; Jay Plasman, University of California, Santa Barbara; Lashawn Richburg-Hayes, R. Hayes & Associates, LLC; Russell Rumberger, University of California, Santa Barbara; and Lori Taylor, Texas A&M University.

The committee also extends its gratitude to members of the staff of the National Academies of Sciences, Engineering, and Medicine, for their significant contributions to this report. Kelly Arrington, senior program assistant, provided key administrative and logistical support, made sure that committee meetings ran smoothly, and provided critical support in managing the manuscript. Constance Citro, former director of the Committee on National Statistics and now its senior scholar, leant to this project her vast knowledge about federal agencies and the data they maintain. Natalie Nielsen, former acting director of the Board on Testing and Assessment, was instrumental in making this project a reality, guiding it from its initial inception to this final report. Kirsten Sampson Snyder and Yvonne Wise masterfully shepherded the report through the review and production process, and Eugenia Grohman provided her always-sage editorial advice.

This Consensus Study Report was reviewed in draft form by individuals chosen for their diverse perspectives and technical expertise. The purpose of this independent review is to provide candid and critical comments that will assist the National Academies of Sciences, Engineering, and Medicine in making each published report as sound as possible and to ensure that it meets the institutional standards for quality, objectivity, evidence, and responsiveness to the study charge. The review comments and draft manuscript remain confidential to protect the integrity of the deliberative process.

We thank the following individuals for their review of this report: Alice Merner Agogino, Mechanical Engineering and Development Engineering Graduate Group, University of California, Berkeley; Dianne Chong, Assembly, Factory & Support Technology (retired), Boeing Research and Technology; Jamel K. Donnor, Holmes Scholars Program and Interdisciplinary Educational Studies Minor, College of William and Mary; Edward H. Haertel, School of Education, Stanford University; Kristen Harper, Policy Development, Child Trends; John Hattie, Graduate School of Education, University of Melbourne, Australia; Brian W. Junker, Department of Statistics, Carnegie Mellon University; Jennifer O'Day, Institute Fellow, American Institutes for Research; Ricki Price-Baugh, Director of Academic Achievement, Council of the Great City Schools; and Deborah J. Stipek, Graduate School of Education, Stanford University.

Although the reviewers listed above provided many constructive comments and suggestions, they were not asked to endorse the conclusions or recommendations of this report nor did they see the final draft before its release. The review of this report was overseen by Lauress L. Wise, Education Policy Impact Center, HumRRO (retired), and Deb A. Niemeier, Department of Civil and Environmental Engineering, University of California, Davis. They were responsible for making certain that an independent examination of this report was carried out in accordance with the standards of the National Academies and that all review comments were carefully considered. Responsibility for the final content rests entirely with the authoring committee and the National Academies.

I have been a member of many committees of the National Academies of Sciences, Engineering, and Medicine and chaired a handful. I have been privileged again to work with Judith Koenig, our study director, whose talent and contributions exceed all superlatives that come to mind. As for my fellow panel members, I have never worked with a more capable and inspiring set of colleagues. They gave generously of their time and worked tirelessly to ensure that the final product accurately represents our consensus findings, conclusions, and recommendations. These efforts manifested the panel members' deep dedication to improving equity across the country. Our work has been the most difficult and humbling I have encountered at the National Academies. And no subject has been as important.

Christopher Edley, Jr., *Chair*
Committee on Developing Indicators of
Educational Equity

Contents

Boxes, Figures, and Tables

BOXES

Summary

The Committee on Developing Indicators of Educational Equity was formed to identify key indicators for measuring and monitoring the extent of equity in the nation's K–12 education system. The purpose of such indicators is not to track progress toward aggregate goals, such as that all students graduate high school within 4 years of entering 9th grade, but to identify *differences* in progress toward that goal, *differences* in students' family background and other characteristics, and *differences* in the conditions and structures in the education system that may affect students' education. A carefully chosen set of equity indicators can highlight disparities, provide a way to explore potential causes, and point toward possible improvements.

Enacting change can be challenging, but it is nearly impossible if there is no information about existing problems. Systematically collected indicators can allow valid comparisons of schools, districts, and states across a number of important student outcomes. No one indicator by itself can tell the full story, but taken together, a set of indicators can provide a detailed and nuanced picture that can inform and enlighten policy makers, policy implementers, state school boards and superintendents, educators, and researchers.

Educational attainment, including, at a minimum, high school completion and a postsecondary credential, is a valued goal for all children in the United States. A high-quality education is in the best interests not only of every individual, but also of society. Failing to attain at least a high school education leaves individuals vulnerable to adverse consequences in adulthood, including a higher likelihood of unemployment, low-wage employment, poor health, and involvement with the criminal justice system. Those adverse adult outcomes for poorly educated individuals have significant costs for the nation as a whole (**Conclusion 1-1**).

Disparities in educational attainment among population groups have characterized the United States throughout its history. Students from families that are white, have relatively high incomes, and are proficient in English have tended to have higher rates of educational attainment than other students, yet they now represent a decreasing proportion of the student population, while groups that have been historically disadvantaged represent an increasing proportion of the student population. An educational system that benefits certain groups over others misses out on the talent of the full population of students. It is a loss both for the students who are excluded and for society (**Conclusion 1-2**).

The history of constitutional amendments, U.S. Supreme Court decisions, and federal, state, and local legislation and policies indicates: (1) a recognition that population groups—such as racial and ethnic minorities, children living in low-income families, children who are not proficient in English, and children with disabilities—have experienced significant barriers to educational attainment; and (2) an expressed intent to remove barriers to education for all students. Educational equity requires that educational opportunity be calibrated to need, which may include additional and tailored resources and supports to create conditions of true educational opportunity (**Conclusion 1-3**). This idea of equity is different from equality, which connotes the idea that certain goods and services are distributed evenly, irrespective of individual needs or assets.

The circumstances in which students live affect their academic engagement, academic progress, and educational attainment in important ways. If narrowing disparities in student outcomes is an imperative, schools cannot shirk the challenges arising from context. Neither can they confront these challenges on their own. Contextual factors that bear on learning range from food and housing insecurity to exposure to violence, unsafe neighborhoods, and adverse childhood experiences to exposure to environmental toxins. Children also differ in their individual responses to stress. Addressing student needs, in light of their life circumstances, requires a wide variety of resources. It is a responsibility that needs to be shared by schools, school systems, other agencies serving children and families, and nongovernmental community organizations (**Conclusion 3-1**).

PROPOSED INDICATORS

The committee emphasizes that an indicator is a measure (e.g., a statistic) that is used to track progress toward objectives or monitor conditions over time. For education, an indicator would allow meaningful examination of equity between key population groups, such as those defined by socio-economic status, race and ethnicity, or English proficiency.

To be useful to policy makers, educators, and other stakeholders, two types of equity indicators are needed: (1) indicators that measure disparities in students' academic achievement and attainment outcomes and engagement in schooling; and (2) indicators that measure equitable access to resources and opportunities, including the structural aspects of school systems that may impact opportunity and exacerbate existing disparities in family and community contexts and contribute to unequal outcomes for students (**Conclusion 2-1**).

To ensure that the pursuit of equity encompasses both the goals to which the nation aspires for its children and the mechanisms to attain those goals, a system of educational equity indicators should balance breadth of coverage with specificity to the appropriate stages of child development and to relevant groups facing disparity. It should also balance consistency across time and place with sensitivity to temporal and geographic context. The committee identified the following characteristics as crucial to striking this balance.

A system of educational equity indicators should (**Conclusion 2-2**):

1. measure multiple dimensions of educational outcomes and opportunities, including changes over time;
2. focus on disparities between the population subgroups most salient for policy attention;
3. use measures that are comparable across time and place, and useful at several organizational scales (classrooms, schools, districts, states, nation);
4. use indicators and measures appropriate to grade level;
5. measure contextual and structural characteristics of or affecting the educational system, such as racial segregation and concentrated poverty;
6. produce frequent, readily understood, high-level summary statistics, in addition to more nuanced statistics;
7. be based on scientifically sound measures; and
8. incorporate mechanisms for continuous improvement based on research and other developments.

Consistent with these conclusions, the committee identified 16 indicators in seven domains. Domains A, B, and C cover individuals: they focus on equity in key measurable outcomes from preschool through the postsecondary transition. Domain D addresses the broader context for the racial, ethnic, economic, and linguistic segregation that confront education in the United States. Domains E, F, and G cover institutions: they address equitable access to opportunities afforded by the education system that can contribute to—or diminish—group differences in achieving key educational outcomes.

Disparities in Outcomes

The committee proposes a set of seven indicators to measure outcomes that we judged to be critically important milestones for success as students proceed from kindergarten through the postsecondary transition: see Table S-1.

Domain A: Kindergarten Readiness

Early childhood experiences set the stage for later academic success. Broadly speaking, kindergarten readiness is the set of foundational skills, behaviors, and knowledge that enable children to successfully transition into kindergarten and achieve academic success throughout the primary grades. From an equity perspective, monitoring kindergarten readiness is important because large between-group disparities become apparent well before children enter kindergarten and can have lasting effects.

- Indicator 1: Disparities in Academic Readiness
- Indicator 2: Disparities in Self-Regulation and Attention Skills

Domain B: K–12 Learning and Engagement

What students learn and how they perform in school positions them for future success as they progress through the K–12 system and as they pursue postsecondary options. To benefit from instruction, students first have to be at school. The positive relationship between instruction time and learning is well documented. Course performance and test scores are well-documented as reliable and valid indicators of academic learning and progress toward educational attainment. Group differences along these dimensions are problematic because they have been found to predict a wide range of longer-term disparities that can impede students from reaching their full potential.

- Indicator 3: Disparities in Engagement in Schooling
- Indicator 4: Disparities in Performance in Coursework
- Indicator 5: Disparities in Performance on Tests

TABLE S-1 Proposed Indicators of Educational Equity

DOMAIN	INDICATORS	CONSTRUCTS TO MEASURE
A Kindergarten Readiness	1 Disparities in Academic Readiness	Reading/literacy skills Numeracy/math skills
	2 Disparities in Self-Regulation and Attention Skills	Self-regulation skills Attention skills
B K–12 Learning and Engagement	3 Disparities in Engagement in Schooling	Attendance/absenteeism Academic engagement
	4 Disparities in Performance in Coursework	Success in classes Accumulating credits (being on track to graduate) Grades, GPA
	5 Disparities in Performance on Tests	Achievement in reading, math, and science Learning growth in reading, math, and science achievement
C Educational Attainment	6 Disparities in On-Time Graduation	On-time graduation
	7 Disparities in Postsecondary Readiness	Enrollment in college, entry into the workforce, enlistment in the military
D Extent of Racial, Ethnic, and Economic Segregation	8 Disparities in Students' Exposure to Racial, Ethnic, and Economic Segregation	Concentration of poverty in schools Racial segregation within and across schools
E Equitable Access to High-Quality Early Learning Programs	9 Disparities in Access to and Participation in High-Quality Pre-K Programs	Availability of licensed pre-K programs Participation in licensed pre-K programs

continued

TABLE S-1 Continued

DOMAIN	INDICATORS	CONSTRUCTS TO MEASURE
F Equitable Access to High-Quality Curricula and Instruction	10 Disparities in Access to Effective Teaching	Teachers' years of experience Teachers' credentials, certification Racial and ethnic diversity of the teaching force
	11 Disparities in Access to and Enrollment in Rigorous Coursework	Availability and enrollment in advanced, rigorous course work Availability and enrollment in advanced placement, international baccalaureate, and dual enrollment programs Availability and enrollment in gifted and talented programs
	12 Disparities in Curricular Breadth	Availability and enrollment in coursework in the arts, social sciences, sciences, and technology
	13 Disparities in Access to High-Quality Academic Supports	Access to and participation in formalized systems of tutoring or other types of academic supports, including special education services and services for English learners
G Equitable Access to Supportive School and Classroom Environments	14 Disparities in School Climate	Perceptions of safety, academic support, academically focused culture, and teacher-student trust
	15 Disparities in Nonexclusionary Discipline Practices	Out-of-school suspensions and expulsions
	16 Disparities in Nonacademic Supports for Student Success	Supports for emotional, behavioral, mental, and physical health

Domain C: Educational Attainment

Education is a critically important way for individuals to pursue their goals in life. On average, higher levels of educational attainment are associated with higher levels of financial, emotional, and physical well-being. Yet research consistently shows between-group differences in educational attainment related to people's race, ethnicity, and gender.

Given the lifelong benefits that accrue with increasing levels of education, the committee's aspiration is for all students to earn a 2- or 4-year college degree. This goal includes high-school graduation, readiness for postsecondary education, and postsecondary matriculation and completion. Because postsecondary persistence and completion are beyond the scope of this report, our indicators are focused on readiness for the transition to 2- or 4-year postsecondary education.

- **Indicator 6: Disparities in On-Time Graduation**
- **Indicator 7: Disparities in Postsecondary Readiness**

Equitable Access to Resources and Opportunities

Disparities in educational opportunities are important to understand and monitor because, at a minimum, they reinforce, and, at worst, they amplify, disparities in outcomes throughout people's lives. While schools are not the only source of opportunity, they can mirror and even exacerbate societal inequities. Yet even in the face of powerful external influences, the investments the nation makes in preschool and K–12 education can play a crucial role in mitigating them. In an effort to maximize attention to such investments, the committee's proposed set of indicators includes high-leverage focal points that can signal problematic group differences in achieving key educational outcomes or progress toward overcoming identified disparities. Along with these school-based opportunities, the committee includes indicators of the role of segregation and structural inequity.

Domain D: Extent of Racial, Ethnic, and Economic Segregation

Segregation, both economic and racial/ethnic, poses one of the most formidable barriers to educational equity. Under conditions of economic segregation, low-income students disproportionately attend schools with high concentrations of other low-income students. Schools that are marked by concentrated poverty often lack the human, material, and curricular resources to meet the academic and socioemotional needs of their populations. Segregation also brings racial differences in exposure to concentrated poverty, leading to nonwhite students being in schools with higher rates of concentrated poverty than other students. This situation exacerbates racial disparities in educational outcomes.

- **Indicator 8: Disparities in Students' Exposure to Racial, Ethnic, and Economic Segregation**

Domain E: Equitable Access to High-Quality Early Learning Programs

Early childhood education is a strong predictor of kindergarten readiness, and one of the most common and policy-relevant out-of-home experiences that young children have. However, there are sizable differences in the availability of high-quality early learning programs and in enrollment between children from lower-income families, families with parents with lower levels of educational attainment, and families in which the parents are not proficient in English and their more advantaged peers. And that availability gap is compounded by a corresponding disparity in the quality of programs that are available to children from families with different income levels.

- **Indicator 9: Disparities in Access to and Participation in High-Quality Pre-K Programs**

Domain F: Equitable Access to High-Quality Curricula and Instruction

The interaction between students and teachers—through curriculum, coursework, and instruction—is at the heart of education. Students' exposure to a rich and broad curriculum, challenging coursework, and inspired teaching is therefore vital for their learning and development. There is no widespread agreement on which specific elements of curriculum, coursework, and teaching matter for student outcomes, but there is evidence that these core elements are not distributed in an equitable way—in relation to either proportionality or need.

There is widespread agreement that teachers are the most important in-school factor contributing to student outcomes, but the research is not as conclusive about which teacher characteristics are associated with effectiveness. From an equity standpoint, the biggest concern is that teachers with more experience and credentials are currently not distributed equally or equitably among schools with different student populations.

Coursework is another central component of academic progress and attainment. Research has long shown that differences in exposure to challenging courses and instruction contribute to disparities in educational outcomes by race, ethnicity, and socioeconomic status. As such, improving access to high-quality advanced coursework across several disciplines represents a potential lever for reducing group disparities in educational attainment.

Access to a broad curriculum that includes courses in art, geography, history, civics, technology, music, science, world languages, and other subjects is important to help all students become well-rounded individuals.

Excellence in academic programming and resources needs to include not only equitable access to advanced placement courses and other advanced coursework, but also meeting the academic needs of students on the other end of the achievement distribution. The adequacy of formal academic supports for students who are struggling to achieve is at least as important as fair access to enrichment opportunities for high-achieving students.

- **Indicator 10: Disparities in Access to Effective Teaching**
- **Indicator 11: Disparities in Access to and Enrollment in Rigorous Coursework**
- **Indicator 12: Disparities in Curricular Breadth**
- **Indicator 13: Disparities in Access to High-Quality Academic Supports**

Domain G: Equitable Access to Supportive School and Classroom Environments

Students need more than challenging courses and effective teachers to thrive academically. They also need physically and emotionally safe learning environments, with a range of supports that pave the way for them to succeed by addressing their socioemotional and academic needs. Safe, supportive school environments and access to counseling, as well as referral to social services, are especially important for students who experience chronic stressors outside of school that affect their learning and development.

- **Indicator 14: Disparities in School Climate**
- **Indicator 15: Disparities in Nonexclusionary Discipline Practices**
- **Indicator 16: Disparities in Nonacademic Supports for Student Success**

As we note above, the purpose of these proposed indicators is to shed light on differences among students, schools, and their contexts. These indicators could serve an important function to alert the public and policy makers to disparities and suggest avenues for further investigation and policy interventions or changes.

In developing the proposed set of indicators, the committee conducted a broad review of the equity indicators reported by other programs, the data sources they use, the indicators they report, and the strategies and mechanisms they use to communicate with stakeholders. There are many publications of key indicators for K–12 education and, more generally, for child well-being, and most publications link to more detailed underlying data. But none of the publications, including those that focus specifically on between-group disparities, presents a fully developed representation based on a carefully articulated concept of equity that covers all student groups of interest. In addition, the indicators in some reports are based on data sources that cannot support subnational detail.

Overall, existing data collection programs and related publications present a mixed picture with regard to their ability to support the committee's proposed set of K–12 educational equity indicators (**Conclusion 2-3**).

RECOMMENDATIONS

The committee recommends a system or set of indicators that are collected and reported on a regular, sustained basis. The committee concludes that it is critical to develop methods for reporting and tracking the educational equity indicators we propose.

We call for the indicators to be collected on a broad scale across the country with reporting mechanisms designed to regularly and systematically inform stakeholders at the national, state, and local levels about the status of educational equity in the United States. A set of key indicators is intended to bring attention to the current status of U.S. education and allow policy makers and the public to identify disparities, explore the causes of those disparities, and decide on actions to address identified inequities, as well as to monitor progress over time. The system we envision would have the same level of priority as the National Assessment of Educational Progress (NAEP), with annual reports that allow the country to monitor progress in making education more equitable from pre-K to grade 12 to the transition to postsecondary education. Given this aspiration, we do not underestimate the level of effort and national will that will be required. That effort will be needed to assemble the necessary data, conduct analyses and data transformations to generate indicators, and implement, evaluate, and improve a system of indicators on a continuing, regular basis.

RECOMMENDATION 1: The federal government should coordinate with states, school districts, and educational intermediaries to incorporate the committee's proposed 16 indicators of educational equity into their relevant data collection and reporting activities, strategic priorities, and plans to meet the equity aspects of the Every Student Succeeds Act.

RECOMMENDATION 2: To ensure nationwide coverage and comparability, the federal government should work with states, school districts, and educational intermediaries to develop a national system of educational equity indicators. Such a system should be the source of regular reports on the indicators and bring visibility to the long-standing disparities in educational outcomes in the United States and should highlight both where progress is being made and where more progress is needed.

RECOMMENDATION 3: In designing the recommended indicator system, the federal government, in coordination with states, school districts, and educational intermediaries, should take care that the system enables reporting of indicators for historically disadvantaged groups of students and for specific combinations of demographic characteristics, such as race and ethnicity by gender. The system also should have the characteristics of effective systems of educational equity indicators identified by the committee.

We note that the system of indicators we propose focuses on the role the education system should play in addressing academic disparities. Although unaddressed in this report, other child-serving agencies play an equally important role in helping at-risk children. The effects of adversity on a child or adolescent depends not only on individual resilience and natural variations in child development, but also on the child's opportunity for experiences, interventions, and supports that may mitigate or even undo the effects of adversity, both material and psychological. Consequently, learning obstacles resulting from the contexts of children's lives are not student deficits barring success, but student needs in search of appropriate opportunities. Research is needed to increase understanding of how various interventions or opportunities are related to individual student needs that are rooted in context. Consensus-building is needed to create indicators and measures that eventually would be included in a broader equity indicator system. For many student needs, screening and responses can best be provided outside of the school setting. Therefore, an indicator system that encompasses all the domains of opportunity important for equity will need to monitor how well student success is supported by other child-serving agencies.

RECOMMENDATION 4: Governmental and philanthropic funders should work with researchers to develop indicators of the existence and effectiveness of systems of cross-agency integrated services that address context-related impediments to student success, such as trauma and chronic stress created by adversity. The indicators and measures should encompass screening, intervention, and supports delivered not only by school systems, but also by other child-serving agencies.

A concerted effort is needed to create the system of equity indicators. Demonstration projects and early prototypes will help catalyze interest in the system and test its feasibility and usefulness.

RECOMMENDATION 5: Public and private funders should support detailed design and implementation work for a comprehensive set of equity indicators, including an operational prototype. This work should involve: (1) self-selected "early adopter" states and districts; (2) intermediaries, such as the Council of the Great City Schools, the Council of Chief State School Officers, and the National Governors Association; (3) stakeholder representatives; and (4) researchers. This work should focus on cataloguing the available data sources, determining areas of overlap and gaps, and seeking consensus on appropriate paths forward toward expanding the indicator system to a broader set of states and districts.

A system of equity indicators needs input and buy-in from a range of stakeholders. This input is needed to develop a process for producing an informative and coherent set of educational equity indicators, determine their content, and ensure that the results will be understood by users. For these purposes, we believe a governing body is needed to provide governance and implementation. We suggest that one analogous to the National Assessment Governing Board that partners with the National Center for Education Statistics for NAEP could be a useful model.

RECOMMENDATION 6: Public or private funders, or both, should establish an independent entity to govern the committee's proposed educational equity indicators. The responsibilities of this entity would include establishing and maintaining a system of research, evaluation, and development to drive continuous improvement in the indicators, measures of them, reporting and dissemination of results, and the system generally. This entity might be structured like the National Assessment Governing Board and might report on both levels of the various outcomes the committee proposes and equity gaps in those indicators, as the Governing Board currently does with NAEP.

Acting on these recommendations will keep in the public eye a critical goal for the nation: to ensure that all students receive the supports they need to obtain a high-quality education from pre-K through 12th grade. Educating all students is fundamental to the nation's ability to grow and develop and to afford all of its people the opportunity to live full and rewarding lives.

1

Why Indicators of Educational Equity Are Needed

If the ladder of educational opportunity rises high at the door of some youth and scarcely rises at the doors of others, while at the same time formal education is made a prerequisite to occupational and social advancement, then education may become the means, not of eliminating race and class distinctions, but of deepening and solidifying them.

This quote is taken from the report of the Commission on Higher Education, which had been established by President Truman. Issued in 1947, the report is historically significant because it represents the first time that a U.S. president commissioned a panel to analyze the country's system of education, a task typically left to the states, as laid out in the Constitution's 10th Amendment. The report is also significant because of the messages it carried and the sweeping changes it called for. Its recommendations included:

> . . . the doubling of college attendance by 1960; the integration of vocational and liberal education; the extension of free public education through the first 2 years of college for all youth who can profit from such education; the elimination of racial and religious discrimination; revision of the goals of graduate and professional school education to make them effective in training well-rounded persons as well as research specialists and technicians; and the expansion of Federal support for higher education through scholarships, fellowships, and general aid.

The report also called for:[1]

> . . . the establishment of community colleges; the expansion of adult education programs; and the distribution of Federal aid to education in such a manner that the poorer States can bring their educational systems closer to the quality of the wealthier States.

These goals are striking because they could have been written yesterday. In the intervening 70 years, there have been other presidential commissions charged with analyzing and evaluating the state of education in the country (e.g., the National Commission on Excellence in Education in the 1980s and the President's Commission on Excellence in Special Education in 2001). There have been many legislative and policy efforts aimed at removing barriers to opportunity for socially and economically disadvantaged groups and holding states and school systems accountable for the academic progress of all of their students (e.g., Title I of the Elementary and Secondary Education Act in 1965 and the No Child Left Behind Act of 2001). A plethora of constitutional amendments, federal and state laws, and court decisions mark the nation's history of attempts to address educational inequities: see Boxes 1-1, 1-2, and 1-3. Despite these efforts, however, the nation has not met many of the aspirations for education equity laid out more than 60 years ago.

Access to high-quality schooling is still uneven across student groups, and "race and class" distinctions (as in the quote above) remain. In recent years, rising income inequality has increased residential segregation, as families move to places where they can afford the cost of housing, which frequently leads to areas with high concentrations of poverty. Black and Latino children are more likely than white children to live in high-poverty areas.

- The rate of black children living in high-poverty areas in 2016 was about six times higher than that for white children (30% and 5%, respectively). The rate for Latino children (22%) was about four times that for white children (Annie E. Casey Foundation, 2018).
- The rate of children living in poverty in 2016 was about three times higher for black children (34%) than for white children (12%). The rate for Latino children (28%) was more than double that for white children: see Figure 1-1.

And as parental education affects family income, black children (12%) were twice as likely as white children (6%) to live in families in which the head of the household did not have a high school diploma. The rate for Latino

---

[1]Cited by Pell Institute: see http://pellinstitute.org/indicators/.

BOX 1-1
Relevant Constitutional Amendments and Court Decisions

The Bill of Rights (the first 10 amendments to the U.S. Constitution, enacted in 1791, protects Americans against infringement by the federal government of their basic rights, such as due process in the criminal justice system. However, nothing in those amendments prevents states from adopting laws that discriminate against a group. It was not until after the Civil War that the 14th Amendment (1868) imposed on the states the obligation to respect due process, equal protection, and other elements of the Bill of Rights for all people.

Roughly 30 years after the passage of the 14th Amendment, in *Plessy v. Ferguson* (1896), the U.S. Supreme Court held that "separate but equal" education and other facilities, legislated in southern states after the end of Reconstruction in 1877, were constitutional under the 14th Amendment. The decision was a major setback for efforts to improve educational opportunities for African American students in these states, given that the separate education facilities provided to them were decidedly inferior to and starved for resources in comparison with the facilities provided for white students. De facto segregation of neighborhoods and schools in other states often had the same result of an inequitable allocation of taxpayer resources for education, while off-reservation day or boarding schools provided by the federal government for Native American students had many drawbacks, and children with disabilities were often warehoused in institutions that provided substandard education.

Almost 60 years after *Plessy v. Ferguson*, the U.S. Supreme Court reversed itself and in *Brown v. Board of Education of Topeka* (1954) declared state laws establishing separate public schools for black and white students to be unconstitutional. In *Brown v. Board of Education II* (1955), the Court remanded future desegregation cases to lower federal courts and instructed school boards to desegregate schools "with all deliberate speed." These decisions set forth a clear goal but gave considerable leeway to localities in how and how fast to implement it. Desegregation of public schools in all southern states was not achieved until 1970 when President Nixon established task forces in each of seven states to implement desegregation plans. Integration was also resisted in many northern states, where schools reflected the clustering of race and ethnic groups in neighborhoods. After *Brown v. Board of Education*, "white flight" to suburban areas, furthered by discriminatory real estate and banking practices, exacerbated de facto residential and school segregation. (See further discussion of contemporary racial and ethnic segregation in neighborhoods in Chapter 5 and in schools in Chapter 6.)

In another 14th Amendment case invoking the equal protection clause, *Plyler v. Doe* (1982), the court struck down revisions to education laws adopted by Texas in 1975 that would have not only withheld state funds for educating children who were undocumented immigrants, but also authorized school districts to deny them enrollment.

BOX 1-2
State Laws on Compulsory School Attendance

In the 19th and early 20th century there was a gradual extension of state laws compelling school attendance at the elementary and secondary levels, which could be in free public schools supported by taxes, in private schools, or through home schooling. Massachusetts in 1852 became the first state to require school attendance; Mississippi in 1917 became the last state to do so. Before the Civil War, localities generally supported only elementary schools; after the Civil War, they began to build high schools.

States vary in the age ranges for which school attendance is required. The minimum compulsory age ranges from age 5 (8 states and the District of Columbia) to age 8 (2 states), with age 6 the most common minimum (24 states). The maximum compulsory age ranges from age 16 (19 states) to age 18 (20 states and the District of Columbia). While there are various exemptions to these age requirements (including that beginning school can be delayed a year and that graduating high school ends the attendance requirement regardless of age), the importance of attending elementary and secondary school is firmly established in all states.[a] With regard to the grade level at which compulsory attendance begins, most states require localities to offer at least half-day kindergarten, although only 17 states and the District of Columbia require kindergarten attendance; 2 other states require kindergarten programs and attendance in particular school districts.[b] No state at present requires pre-K attendance, although more and more states are requiring districts to provide pre-K programs.

---

[a]See http://www.ncsl.org/documents/educ/ECSCompulsoryAge.pdf.
[b]See https://www.ecs.org/kindergarten-policies/.

BOX 1-3
Federal Laws Affecting Educational Equity

The 1964 Civil Rights Act extended the equal protection of the 14th Amendment in a sweeping manner. In addition to its broad mandates regarding public accommodations engaged in interstate commerce and its prohibition against state and municipal governments denying access to public facilities, it encouraged desegregation of public schools and authorized the U.S. Attorney General to file suits for enforcement.

The 1965 Elementary and Secondary Education Act (ESEA), part of President Johnson's War on Poverty, emphasized reducing disparities in educational achievement by providing resources to states and school districts to improve educational opportunities for children in low-income families (Title I) and improving education from kindergarten through 12th grade in other ways. Title I funds are allocated according to a formula.

The 1968 Bilingual Education Act (Title VII of ESEA) provided competitive grants to school districts for innovative educational programs for children with lim-

BOX 1-3 Continued

ited English-speaking ability. It explicitly recognized that educational equity could require different kinds of instruction for different student populations.

The 1974 Equal Educational Opportunities Act prohibited states from denying equal educational opportunity to students based on gender, race, color, or nationality. Specifically, states could not allow educational institutions to implement intentional segregation, neglect to resolve intentional segregation, force students to attend a school outside their neighborhood that promoted further segregation, discriminate in employing faculty and staff, or fail to remove language barriers that prevented students' equal participation in English classes.

The 1975 Education for All Handicapped Children Act, reauthorized in the 1990 Individuals with Disabilities Education Act, required public schools to educate all children with disabilities by creating an educational plan for them with their parents' input to replicate as closely as possible the educational experience of nondisabled students. It replaced earlier acts that provided grants to states for education of children with disabilities but that did not require such education.

The 1994 Improving America's Schools Act reauthorized ESEA. Notable changes included provisions in Title I to hold schools accountable for the educational achievement of disadvantaged students at the same level as other students and additional resources for bilingual and immigrant education.

The 2001 No Child Left Behind Act (NCLB), which again reauthorized ESEA, increased the federal role in education. It required states to develop or adopt assessments in reading, mathematics, and science and give them to all students at specified grade levels, showing separate results by racial/ethnic group and for students who were economically disadvantaged, had disabilities, or had limited English proficiency. States were required to provide highly qualified teachers to all students. Schools that received Title I funding had to make "adequate yearly progress" (an annual measure of academic growth that is set by each state in collaboration with the U.S. Department of Education); penalties up to closing down and restructuring the school were to be imposed should progress goals not be achieved.

The 2015 Every Student Succeeds Act is the most recent reauthorization of ESEA. This act cut back on the expansion of the federal role in education that was stipulated under NCLB. Under ESSA, states are accountable for focusing resources on low-performing schools and traditionally underserved students who consistently demonstrate low academic performance. The law requires data on student achievement and graduation rates to be reported as well as action in response to that data. States are still required to develop assessment standards, submit their plans to the U.S. Department of Education for approval, and use the assessments for all students in specified grade, but the act delegated back to the states the determination of penalties for poor performance. It also added a requirement that all schools offer college and career counseling and Advanced Placement courses to all students.[a] However, unlike NCLB, states, districts, and schools will determine what support and interventions are implemented. Under ESSA, states have flexibility to chart their own path to educational success, but they must submit a plan to the U.S. Department of Education explaining how they will reach these goals.

---

[a]For further information about NCLB and ESSA see https://www.ed.gov/essa.

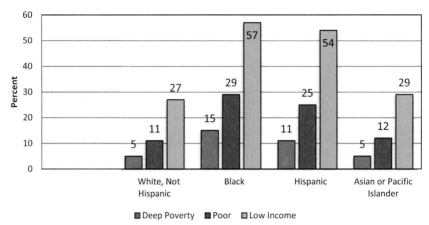

FIGURE 1-1 Percentage of children in economically disadvantaged families by race and ethnicity, 2017.
NOTES: Deep Poverty: Under 50% of Federal Poverty Level (FPL); Poor: Under 100% of FPL; Low Income: Under 200% of FPL.
SOURCE: Data from Child Trends, 2019, see https://www.childtrends.org/?indicators=children-in-poverty.

children (32%) was more than five times that for white children (Annie E. Casey Foundation, 2018).

Most school districts reflect the demographic and socioeconomic composition of their neighborhoods. School assignment policies that send all (or many) children from a high-poverty neighborhood to the same school create schools with high concentrations of children living in poverty. As we document in this report, schools serving children from low-income families tend to have fewer material resources (books, libraries, classrooms, etc.), fewer course offerings, and fewer experienced teachers. The educational opportunities available to students attending these schools are not of the same quality as those in schools in more affluent neighborhoods.

Education is sometimes characterized as the "great equalizer," but to date, the country has not found ways to successfully address the adverse effects of socioeconomic circumstances, prejudice, and discrimination that suppress performance for some groups. The rapidly changing demographic characteristics of the population of school-age children mean new challenges for school systems. Figures 1-2 and 1-3 show some of the differences by race and ethnicity and home language.

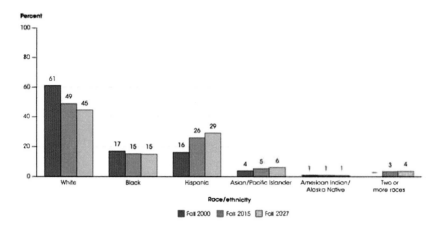

FIGURE 1-2 Percentage distribution of public school students enrolled in pre-kindergarten through grade 12, by race and ethnicity: fall 2000, 2015, and projected for 2027.
NOTES: — Not available. Race categories exclude persons of Hispanic ethnicity. Although rounded numbers are displayed, the figures are based on unrounded estimates. Detail may not sum to totals because of rounding.
SOURCE: Snyder, de Brey, and Dillow (2019, Table 203.50).

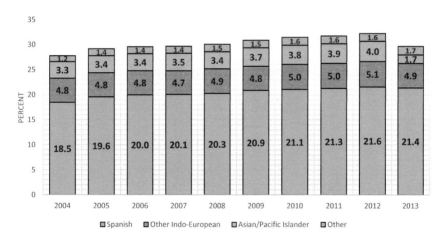

FIGURE 1-3 Percentage of children aged birth to 17 with household languages other than English, 2004-2013.
SOURCE: Data from Child Trends, 2016; see https://www.childtrends.org/indicators/dual-language-learners.

BOX 1-4
Charge to the Committee

[The] committee will develop a set of indicators around equity in educational outcomes and relevant inputs for pre-K through the transition to post-secondary education. To develop the indicators, the committee will examine existing indicator systems in education and child well-being. The committee will review a wide range of research related to these systems and the types of outcomes that are important for the education system to achieve (e.g., readiness for the next level of schooling, opportunities to learn, academic performance, persistence, and engagement). The committee will also examine research on school and non-school inputs that are related to those outcomes, the extent of inequality in these inputs and outcomes, and interventions that have been shown to improve outcomes. The committee will prepare a report that:

- presents the indicators that it recommends for use in education;
- uses the available data on those indicators to describe the nation's status in terms of improving such outcomes as achievement, graduation rates, college and career readiness, and postsecondary educational attainment for students whose characteristics or circumstances put them at risk of school failure;
- highlights potentially promising points of intervention for policymakers;
- identifies gaps in existing data and research; and
- identifies future directions for monitoring and improving progress on the indicators, including the major considerations in developing such a system and sustaining it over time.

GOALS OF THIS REPORT

This report documents the work and recommendations of the 15-member Committee on Developing Indicators of Educational Equity. The committee's charge is shown in Box 1-4.[2] In accordance with this charge, the committee identified a set of key indicators that would measure the extent of disparities in the nation's elementary and secondary education system. Their purpose would not be to track progress toward an aspirational goal measured in the aggregate per se, such as that all students graduate high school within 4 years of entering ninth grade, but to track and shed light on group *differences* in progress toward that goal, *differences* in students' family background and other characteristics, and *differences* in the conditions and structures in the education system

[2]The American Educational Research Association, Atlantic Philanthropies, the Ford Foundation, the Spencer Foundation, the U.S. Department of Education, the William T. Grant Foundation, and the W.K. Kellogg Foundation provided support for this work.

that can exacerbate or mitigate the effects of those characteristics. Such a system of indicators could serve an important function to alert the public and policy makers to consequential disparities and suggest avenues for further investigation and policy formulation.

The committee recommends developing a system of key equity indicators that would be collected and reported on a regular, sustained basis. We call for the system to operate on a large scale with reporting mechanisms designed to regularly and systematically inform stakeholders at the national, state, and local level about the status of educational equity in the United States. The system we envision would rise to the same level of priority as the National Assessment of Educational Progress (NAEP) with annual reports that allow the country to monitor progress in making education more equitable.

As most readers of this report know, the literature base on educational equity is immense. There are multitudes of studies that document disparities based on race, ethnicity, parental education, family income, native language, and other group definitions; numerous data collections that are available for use in these studies; and uncountable numbers of evaluations and recommendations. The present report is certainly not the first to suggest a set of indicators to monitor education equity.

Given the breadth and depth of information that is already available, why is another report needed? What does our report contribute to the field? In the five chapters that follow, we document what is known about inequities in education and the factors that contribute to them. Much of what we say is not new or ground-breaking. Many of the indicators we propose have long formed the bedrock of attempts to measure educational achievement and attainment of education credentials. However, the committee hopes that this report, anchored in the most recent research, will bring a new and heightened level of attention to educational equity. The indicators we suggest document consequential disparities that have the potential to help policy makers, parents, and others improve education policy or practice and support both formal and informal evaluations of effectiveness. To this end:

- We propose a manageable number of key indicators that cover the full range of pre-K to grade 12 education.
- Our recommendations reflect consensus among experts from diverse areas of expertise.
- Our recommendations acknowledge and emphasize that contextual factors, including segregation, are important for education.
- We recommend high-level structures that could be used to implement our proposed indicators.

COMMITTEE'S CONCEPTION OF EQUITY

In everyday conversation, the terms *equity* and *equality* are often used interchangeably. In technical contexts, their meanings differ in important ways. Equality generally connotes the idea that goods and services are distributed evenly, that is, everyone gets the same amounts, irrespective of individual needs or assets. The "starting point" is irrelevant—including the endowments—both positive and negative—that each individual brings to a situation. In contrast, equity incorporates the idea of need. The idea of need replaces a mechanistic approach to equality in which everyone receives the same amount of whatever is being distributed. Indeed, equity means that distribution of certain goods and services is purposefully unequal: for example, the most underserved students may receive more of certain resources, often to compensate or make up for their different starting points. For other terms that are important to the committee's work, see Box 1-5.

The committee's specific purpose is to develop indicators that document consequential disparities and thus offer insights to policy makers, policy implementers, state school boards, superintendents, educators, researchers, and others to help improve education policy and practice and also to support both formal and informal evaluations of effectiveness. For this pur-

---

BOX 1-5
Definitions of Terms

Between-group difference: a numerical difference between two or more population groups in any given measure (e.g., test scores, percent proficient, years of education); an objective description of differences without any judgments about their magnitude, significance, or impacts.

Disparity: a between-group difference that the committee believes matters in terms of educational outcomes.

Equality: the idea that goods and services are distributed evenly (i.e., everyone gets the same amounts), irrespective of individual needs or assets; the "starting point" is irrelevant—including the endowments that an individual brings to a situation, both positive and negative.

Equity: the idea that need replaces a mechanistic approach to equality (see above); that the distribution of certain goods and services is purposefully unequal so that the neediest of students may receive more of certain resources, often to compensate or make up for their different starting points.

Indicator: a measure, or statistic, used to track progress toward objectives or to monitor the health of an economic, environmental, social, or cultural condition over time.

Inequality: situations in which goods and services are not distributed evenly.

Inequity: situations in which differences in need are not adequately considered.

pose, there is a need for indicators of disparities in key student outcomes related to educational achievement and attainment of credentials and of access to educational resources (e.g., effective teachers and high-quality curricula). Indicators also need to address disparities in access to opportunities to address structural disadvantages. Accordingly, the committee's working definition of educational inequity is three-pronged. There is an inequity when:

- there is an excessive disparity between groups with respect to an educational outcome, such as high school graduation or access to a resource;
- there is an unacceptably poor fit between resources and student needs; or
- there is inadequate effort to mitigate the effects of deleterious segregation or some structural disadvantage faced by a group of learners, such as students from financially disadvantaged families, blacks, Latinos, or English learners.

A great many circumstances qualify as inequitable under this three-pronged definition. Moreover, stakeholders, including parents and students, may have different assessments of success in reducing inequities and, indeed, what is meant by an outcome such as a "high-quality education." Furthermore, there is no scientific basis for specifying a particular value for "excessive" or "unacceptable" or "inadequate." When a set of indicators is identified and becomes a factor in making decisions, people using the indicators—both policy makers and those affected by policy—will need to make their own judgments as to the importance of an observed disparity and what can be done to reduce it, considering other calls on political will and resources.

In this report, we do not specifically address how much inequity is too much or what action level may be appropriate for every actor in the complex, decentralized U.S. education system. However, in our recommendation of measures for inclusion in an education equity indicator system, we do try to focus on which education inequities matter most, based on the historical record of policy preferences revealed in legislation and court decisions and research findings on the consequences for student success in later life.

GUIDING CONCLUSIONS

The committee's review of research on between-group differences in educational attainment (such as obtaining a diploma or other credential), the effects of these differences, and the legislation targeted at reducing these differences led us to draw three conclusions that guided our work.

CONCLUSION 1-1: Educational attainment, including, at a minimum, high school completion and a postsecondary credential, is a valued goal for all children in the United States. A high-quality education is in the best interests not only of the individual, but also of society. Failing to complete at least a high school education leaves individuals vulnerable to adverse consequences in adulthood, including a higher likelihood of unemployment, low-wage employment, poor health, and involvement with the criminal justice system. Those adverse adult outcomes for poorly educated individuals have significant costs for the nation as a whole.

CONCLUSION 1-2: Disparities in educational attainment among population groups have characterized the United States throughout its history. Students from families that are white, have relatively high incomes, and are English proficient have tended to have higher rates of educational attainment than other students, yet they now represent a decreasing proportion of the student population, while groups that have been historically disadvantaged represent an increasing proportion of the student population. An educational system that benefits certain groups over others misses out on the talent of the full population of students. It is a loss both for the students who are excluded and for society.

CONCLUSION 1-3: The history of constitutional amendments, U.S. Supreme Court decisions, and federal, state, and local legislation and policies indicates: (1) a recognition that population groups—such as racial and ethnic minorities, children living in low-income families, children who are not proficient in English, and children with disabilities—have experienced significant barriers to educational attainment; and (2) an expressed intent to remove barriers to education for all students. Educational equity requires that educational opportunity be calibrated to need, which may include additional and tailored resources and supports to create conditions of true educational opportunity.

COMMITTEE'S CONCEPTION OF AN INDICATOR SYSTEM

The value of an indicator system is that it brings attention to existing conditions, allows one to identify problems, provides a way to explore potential causes of those problems, and points toward actions to alleviate the problems. Equity indicators, if collected systematically and well, can allow valid comparisons of schools, districts, and states across a number of important student outcomes and resources. No one indicator by itself can tell the full story, but taken together, the set of indicators can provide a detailed and nuanced picture that speaks to policy makers and other stakeholders about students' educational status. Enacting change can be challenging, but it is nearly impossible if there is no information about existing problems.

What Is an Indicator?

An indicator is a measure used to track progress toward objectives or to monitor the health of an economic, environmental, social, or cultural condition over time. Different sorts of measures are used in different contexts. For example, unemployment rates, infant mortality rates, and air quality indexes are all indicators. In the field of education, school districts typically administer standardized reading assessments at specific grades to monitor how well students are meeting basic benchmarks in reading. Other commonly used education indicators include high school graduation rates, rates of truancy, ratios of teachers to students, and per-pupil expenditures, as well as measures of less quantifiable factors, such as teachers' and students' attitudes.

Indicators—literally, signals of the state of whatever is being measured—can cover outcomes, the presence or state of particular conditions, or the effectiveness of management approaches. They can be used to measure change over time or for comparisons among outcomes, conditions, or measures of effectiveness in different places. Although indicators are usually quantitative, they can either be straightforward measures of a single phenomenon, such as the number or percentage of students who graduate in a given year, or composite measures. A composite indicator is a measure of a more complex phenomenon, such as college readiness, and may incorporate a number of variables that capture aspects of what is being measured. Thus, an indicator is not the same thing as a statistic. As a primer on education indicators explained, statistics "need context, purpose, and meaning if they are going to be considered" indicators (Planty and Carlson, 2010, cited in National Research Council, 2012, p. 4).

Example from Economics: Monthly Jobs Report

Each month the American public receives a summary of the nation's employment situation.[3] Produced by the U.S. Bureau of Labor Statistics (BLS), the summary provides key information for the current and prior month on unemployment and labor force participation, jobs, hours, and earnings. It provides unemployment rates for adult men, women, and teenagers, by race and ethnicity and by educational attainment, and job gains and losses for about two dozen sectors (e.g., manufacturing, health care).

---

[3]See https://www.bls.gov/news.release/empsit.toc.htm: text and Tables A and B.

These key indicators, which are eagerly awaited by the media, Congress, the executive branch, and the private sector, are only a few in a vast cornucopia of employment and unemployment data that provide more detail by worker and job characteristics and geographic areas. The summary's role is to furnish a small set of high-quality, objective, and timely indicators of whether employment conditions are getting better or worse, overall and for key sectors. The summary often triggers further exploration of the detailed data that may, in turn, lead to policy actions, such as extending federal unemployment benefits to workers in high unemployment states.

A Possible Example from Education: Yearly Educational Equity Report

There are currently no indicators for monitoring the status of educational equity in the same way as the BLS summary does for the employment situation in the country. Yet educational equity is an equally important measure of the nation's well-being. We draw from the labor statistics field to explain how a system of indicators would work in an educational equity context. The BLS provides a useful example because its indicators are succinct, familiar, and consistently reported. Understandably, of course, it is not an exact model of what is needed for educational equity. For educational equity, one needs to know about opportunity, and indicators need to be multidimensional, provide information about different grade levels, and meet the needs of thousands of school districts and schools.

A great many indicators and supporting data are regularly produced, not only by federal agencies, but also by state and local agencies and nongovernmental organizations. Examples include the longstanding NAEP, the congressionally mandated annual *Condition of Education* of the National Center for Education Statistics (NCES),[4] which is backed up by the extensive data in the annual *Digest of Education Statistics*,[5] and the education section in the annual *Kids Count Data Book: State Trends in Child Well-Being* of the Annie E. Casey Foundation[6] (see Appendix A). Indeed, the nation's interest in education indicators predates its interest in labor force indicators: the predecessor to NCES was established in 1867, while the predecessor to BLS was established in 1884.

However, unlike the unemployment rate and some other federal economic indicators, none of the available education indicators is officially designated as a "key" indicator,[7] and there is no corresponding regularly

---

[4]See https://nces.ed.gov/programs/coe/.

[5]See https://nces.ed.gov/programs/digest/.

[6]See https://datacenter.kidscount.org/publications/.

[7]See https://www.whitehouse.gov/sites/whitehouse.gov/files/omb/assets/OMB/inforeg/statpolicy/dir_3_fr_09251985.pdf.

published, eagerly awaited, widely publicized report, with supporting data, analogous to the BLS *Employment Situation Summary*. Because education is a key predictor of economic success in life and is included as such in the BLS *Employment Situation Summary*, it would seem worthwhile to have key education indicators with similar stature. Similarly, just as the *Employment Situation Summary* includes indicators for population groups, defined by race, ethnicity, and gender, it would seem worthwhile to have key education indicators that address groups, given the disparities that have long plagued the U.S. education system.

An "Educational Equity Summary" could usefully highlight disparities between key populations groups, such as those defined by race and ethnicity, family income, and other dimensions of public and policy interest on two critical sets of indicators: (1) educational outcomes, such as participation, achievement, behaviors, and attainment; and (2) opportunities provided by the education system, such as access to effective teachers or to high-quality preschool programs. Such information would help target interventions, research, and policy initiatives to reduce disparities. Reports such as the Annie E. Casey Foundation's *Race for Results*[8] and NCES's *Status and Trends in the Education of Racial/Ethnic Groups*[9] are helpful, but they are not regularly published and do not cover the range of indicators that are needed to track educational equity and inform policy.

COMMITTEE'S APPROACH TO ITS CHARGE

The committee conducted a broad review of evidence related to educational equity and educational equity indicators, including:

- the types of positive outcomes that are important for the education system to achieve (e.g., readiness for the next level of schooling, opportunities to learn, academic performance, and engagement in school) from pre-kindergarten through the transition to post-secondary education or other rewarding pursuits;
- school and nonschool inputs and conditions that influence those outcomes;
- the extent of disparities in outcomes and in relevant school inputs;
- interventions that have been shown to improve outcomes; and
- interventions for which, even if evidence is minimal, there is a strong theoretical basis as judged by recognized experts.

---

[8]See http://www.aecf.org/resources/2017-race-for-results/.
[9]See https://nces.ed.gov/programs/raceindicators/.

Two aspects of our review were an assessment of existing data systems of potential use for indicators of educational equity and an assessment of relevant publications: see Appendixes A and B, respectively. In addition, we reviewed the data and methodological opportunities and challenges for developing K–12 educational equity indicators: see Appendix C.

The committee held five in-person meetings and numerous tele-conferences. The first two meetings included open session time to speak with stakeholders and other groups that are involved with collecting data and reporting education indicators. The committee also commissioned a set of authors to conduct literature reviews to help evaluate the empirical basis for potential indicators. A subset of these authors participated in the second meeting and suggested ways to structure the analyses of the research base and recommended other potential authors. The agendas for both meetings are in Appendix D.

ORGANIZATION OF THE REPORT

Five chapters follow this introduction. Chapter 2 discusses the purposes of a system of indicators of educational equity and explains our approach to identifying indicators. Chapter 3 discusses family, home, and neighborhood contextual factors as they relate to educational equity.

Chapters 4 and 5 present our suggestions for indicators. In Chapter 4, we discuss indicators of disparities in student outcomes, and in Chapter 5, we discuss indicators of disparities in access to important resources and opportunities

Chapter 6 presents the committee's recommendations for implementation of an indicator system, addressed to the key audiences for this report. Some useful indicators are ready for prime time for the nation, states, school districts, and schools, while others are ready for some but not all levels of aggregation, and still others require additional research and development. Recommendations also address paths forward from the report to a useful and used system of educational equity indicators.

Appendixes A and B detail the committee's assessments of existing data and indicators for potential use in our recommended system and of relevant publications. Appendix C discusses data and methodological issues. Appendix D contains the agendas for the committee's two public meetings. Appendix E provides biographical sketches of the committee members and staff for this project.

2

Committee's Framework for Indicators of Educational Equity

Indicators are intended to focus attention on a few, key, readily interpretable facts that tell users about the current status of a topic or field and highlight areas in need of improvement. For the purposes of educational equity, indicators should convey—in an easily understood way—the range, nature, and magnitude of disparities, as well as measures of any narrowing or widening over time. Indicators might be useful for many purposes, including research or accountability. The committee, however, was most attentive to information that could inform efforts to improve policy and practice.

Currently, equity is a prominent focus for education policy makers and, in turn, for those who implement policy; this is in part because equity was a major theme in the *Every Student Succeeds Act* (ESSA, 2015) and its predecessor *No Child Left Behind* (NCLB, 2002). States are working to develop and incorporate equity indicators in their required plans, while the federal government is working to review and approve state plans. As stated by one of the individuals we interviewed, "[Nearly] every national level organization related to education is having a discussion about equity. They are including equity in their priorities, core principles, strategic plans, and member activities." A set of research-based indicators could help to ground the related but disparate conversations that are taking place and provide a common starting point for many stakeholders in pursuing their own missions. Another interviewee said that the committee's indicators would serve as a "North Star" to guide their process for exploring, investigating, and studying disparities.

This chapter begins by summarizing some of the information about indicator systems we gleaned from: (1) expert guidance on indicator systems; (2) existing programs and initiatives focused on educational equity; and (3) stakeholder insights on the purposes and uses of indicators. It then describes our own framework, or architecture, for a comprehensive set of key equity indicators.

EXPERT GUIDANCE

There are numerous publications that offer guidance on how to set up an indicator system. We primarily relied on the six, listed below in chronological order, written by authorities in the field of education indicators:

1. Jeanie Oakes (1986), *Education Indicators: A Guide for Policy Makers*
2. Shavelson et al. (1987), *Indicator Systems for Monitoring Science and Mathematics Education* and the accompanying sourcebook
3. Marshall Smith (1988), *Education Indicators*
4. Field, Kuczera, and Pont (1988), *Ten Steps to Equity in Education*
5. Bryk and Hermanson (1993), *Educational Indicator Systems: Observations on Their Structure, Interpretation, and Use*
6. Planty and Carlson (2010), *Understanding Education Indicators*

These writings discuss different types of indicators (based on a single statistic or a compound statistic) and different uses for indicators (to report on status, monitor change, project future patterns). They offer advice on ways to design a system, and they provide criteria for evaluating the technical qualities essential for the indicators included in the system. They also provide guidance on the process of developing both the conceptual and operational definitions of indicators, determining how to measure the construct, and collecting the needed data.

EXISTING INITIATIVES

Indicators have long been used in the education arena—one of the earliest was the National Assessment of Educational Progress (NAEP).[1] Designed to track educational achievement over time, NAEP has reported reading and mathematics achievement results for students aged 9, 13, and 17 since 1971. Although not usually referred to as a system of equity indicators, NAEP has highlighted the achievement gaps among the nation's students. Routinely published reports track performance differences for students grouped by race, ethnicity, and sex, and provide disaggregated results for students with disabilities, English learners, and students who receive free

---

[1]For a review of milestones in NAEP's development, see Box 6-1 in Chapter 6.

and reduced-price school lunch. NAEP maintains a sophisticated website with a user-friendly dashboard that allows users to select the information they want to see. Over the years, NAEP has evolved and adapted its methods to meet various policy needs. NAEP has expanded the grade levels included, the subject matter tested, and the reporting levels, first to the state level and then to the urban district level. It is difficult to overestimate the impact NAEP has had on education in the country, particularly the role it has played in raising awareness about disparities in achievement. These impacts are well documented (Bourque, 2009; Casserly et al., 2011; Glaser, Linn, and Bohrnstedt, 1997; Jones and Olkin, 2004; NCES, 2012).

There are numerous other initiatives that have contributed to understanding of disparities in education. Government agencies, such as the National Center for Education Statistics (NCES) and the Census Bureau, collect an enormous amount of data that have been used to evaluate educational equity. For example, NCES annually prepares the *Condition of Education* report that contains indicators on the state of education in the United States, from pre-kindergarten through postsecondary education, as well as labor force outcomes and international comparisons. The data for these indicators are obtained from many different providers—including students and teachers, state education agencies, local elementary and secondary schools, and colleges and universities—using surveys and compilations of administrative records. These data also become part of the *Education Digest* published by NCES and are easily available and widely used.

In addition to the *Condition of Education*, NCES uses these data to prepare other reports, of which the most relevant for our purposes is a series of annual publications called *Status and Trends in the Education of Racial and Ethnic Groups*. The data are also used by others. The Annie E. Casey Foundation draws from these data to prepare its *Kids Count* reports; more recently, the foundation uses the data to monitor and evaluate educational equity in a series of reports, *Race for Results: Building a Path to Opportunity for All Children*. Another user is the Federal Interagency Forum on Child and Family Statistics, which annually publishes *America's Children: Key National Indicators of Well-Being*. Child Trends draws from these data to create short easily digested policy briefs.

Another important data resource is the Civil Rights Data Collection (CRDC) program of the Office for Civil Rights (OCR) of the U.S. Department of Education. Since 1968, a wide variety of data have been collected on key education and civil rights issues in public school, including information about student enrollment and educational programs and services, most of which are disaggregated by race and ethnicity, sex, English proficiency, and disability. "The CRDC is a longstanding and important aspect of the [Office for Civil Rights'] overall strategy for administering and enforcing the civil rights statutes for which it is responsible."[2] OCR prepares "first look" briefs and focused reports that are easily downloaded from its website. The topics of these briefs differ from year to year. The two briefs issued in April 2018 based on data from the 2015-2016 data collection are the *STEM* [science, technology, engineering, and math] *Course Taking Issue Brief* and the *School Climate and Safety Issue Brief*. The CRDC also provides ready access through a search feature to three special reports for school districts and schools: *English Learner Report, Discipline Report,* and *Educational Equity Report*: they are provided in Excel spreadsheets. The search feature allows users to generate state, district, and school reports of user-selected data elements disaggregated by user-specified demographics.

The Committee's Challenge: Improving the Current Situation

Given that a variety of equity indicators are already reported, we considered the ways that our efforts might improve the situation. In what ways do the existing initiatives fall short? What can we suggest that would fill existing gaps or make the whole (collection of indicators) greater than the sum of its parts?

To answer these questions, we first conducted a broad review of the equity indicators reported by other organizations. We considered efforts by both government agencies and nongovernmental organizations (NGOs), and we looked at initiatives that targeted equity from the outset as their primary purpose (e.g., CRDC, Annie E. Casey Foundation's *Race for Results* project), as well as those that just touch the periphery of equity, such as by reporting disaggregated results. We considered 19 of the organizations that produce various reports and briefs intended for a wide spectrum of audiences. Some are involved in all the steps of producing indicator reports (e.g., National Center for Education Statistics), from collecting data to reporting the results. Others make use of data collected by government agencies to develop their own indicators and associated reports (e.g., Child Trends). Still others make use of indicators developed by others to include in their own reports. Some organizations publish reports on a regular basis, most often annually (e.g., *Kids Count,* from the Annie E. Casey Foundation); others publish briefs when the findings warrant (e.g., Child Trends, CRDC). Appendix B provides details about these 19 existing initiatives. We selected reports from seven organizations to explore in depth—to learn more about the indicators that are reported, the data they are derived from, how and when they are reported, and how they are intended to be used and by whom. The reports include:

---

[2]See https://www2.ed.gov/about/offices/list/ocr/data.html?src=rt.

1. Annie E. Casey Foundation: *Kids Count Data Book* and *Race for Results*
2. Council of the Great City Schools: *Academic Key Performance Indicators: 2018 Report*
3. Education Law Center and Rutgers University: *Is School Funding Fair? A National Report Card on Funding Fairness*
4. Federal Interagency Forum on Child and Family Statistics: *America's Children: Key National Indicators of Well-Being*
5. National Center for Education Statistics: *Condition of Education* and *Status and Trends in the Education of Racial and Ethnic Groups*
6. National Institute for Early Education Research, Rutgers University: *State of Preschool* yearbooks
7. U.S. Department of Education, Office of Civil Rights: Civil Rights Data Collection *First Look Issue Briefs* and *Special Reports*

In a related vein, we considered whether an equity indicator system has already been created, in effect, by the data collection and "report card" obligations states face under the federal Every Student Succeeds Act of 2015 (ESSA).[3] Because federal impositions on states are politically problematic, the statutory reporting requirements are loosely defined and cover a very limited domain of education inputs and outcomes, at least when compared with the range of variables discussed in the research literature (see National Urban League, 2019). Moreover, state indicator systems under ESSA continue to evolve, have varied designs, and produce data in ways that may not be comparable with each other or even wholly consistent within a state. Finally, the federal requirements not only are underspecified in a technical sense, but also are only the minimum requirements; states and districts are free to augment these base indicator systems.

STAKEHOLDER INSIGHTS ON USES

To gather information about ways equity indicators might be used, the committee heard presentations during its first two meetings (see agendas in Appendix D) and solicited informal input from representatives of a small set of government agencies and organizations that provide information, support, and analytic assistance to the education policy community. This group included the following:

- representatives from various member organizations that work on behalf of cities, states, school boards, and school administrators, such as the Council of Chief State School Officers, the Council of

---

[3]See https://www2.ed.gov/policy/elsec/leg/essa/essastatereportcard.pdf.

the Great City Schools, the National Governors Association, and the national Parent Teacher Association (PTA);

- independent organizations, such as the Education Trust and the Alliance for Excellent Education;
- organizations that have developed related indicator and reporting systems, such as the Annie E. Casey Foundation and Child Trends; and
- government agencies that collect relevant data, such as the CRDC, the Census Bureau, and NCES.

As noted above, equity is a prominent focus for education policy makers and, in turn, for those who implement policy. Several individuals talked about the need to understand disparities and determine actions to take. States, districts, and other entities may be aware of inequities in their school systems, but they need help in judging their magnitude and sources and in identifying strategies for addressing them. These groups would benefit from clear, understandable metrics that highlight problem areas and could be used in conversations with various constituencies. They would be seeking ways to measure, define, and track their progress in closing gaps and would especially value information that helps them to make research-based decisions about investments and strategies that are likely to increase educational equity over the long term. Figure 2-1 summarizes the information we learned from our information gathering.

FIGURE 2-1 Potential uses of educational equity indicators by different stakeholders.

State education agencies are among the most likely organizations to adopt or use information from the committee's proposed indicators. The existing data collection and reporting burdens on states are a major consideration for any data-related effort, and states (and school districts) would be more likely to adopt indicators of educational equity that dovetail with how they are thinking about equity and the types of data they already collect as part of their own accountability and monitoring efforts, and if they believe they can learn something about themselves from the indicators. Given the current focus on equity in the country, key intermediaries can use the indicators to start and advance conversations about educational equity with states.

Another challenge identified by some stakeholders relates to data that are applicable across state lines. For example, it can be difficult to compile academic indicators in grades 3-8 across state lines because there are not many common data points that illustrate whether states or districts are doing well academically. As one respondent said, "You would think it would be easy with all the testing that we do. But the data that you might collect doesn't mean the same thing across state lines."

Finally, some individuals with whom the committee consulted noted that the design of an indicator system should avoid making the perfect the enemy of the useful; an exhaustive, research-based system would likely be overwhelming and unwieldy for decision makers and those who would implement a system. Thus, the committee reasoned, attention to parsimony—while challenging in a research-driven exercise—would be a valuable contribution to the cacophonous discussion of how best to define and gauge educational equity.

CONSIDERATIONS IN DETERMINING KEY EQUITY INDICATORS

To be effective, a system of equity indicators should provide information that users view as important, credible, and valuable. The system should include indicators that represent constructs that are malleable (capable of being changed) and actionable (easily translated into a plan of action). They should be amenable to change as a consequence of educational policy or practice interventions, and this relationship should be backed by empirical research. Some indicators can play a descriptive, signaling role by calling attention to significant disparities in resources and learning opportunities, such as the distribution of school suspensions and enrollment in advanced placement courses by race and ethnicity across schools and over time. Indicators are much more powerful if the conditions they measure can be shown to be consequential for valued outcomes, such as high school completion and successful transitions to postsecondary education.

Our review of existing initiatives revealed that much of the needed data are available. There is, in fact, a wealth of data on pre-K to grade 12 education and beyond and a large number of education indicator reports prepared by government and nongovernment agencies. However, these existing data and reports are not sufficient for the system of educational equity indicators as we have conceptualized it. One set of problems pertains to the ways the data are collected and stored, in comparison with what our proposed system will require. The data are scattered across different databases, collected through different sampling procedures of different populations, based on data collections from different years and administered at different intervals, and varied across jurisdictions in their technical specifications. Another important set of issues concerns the extent to which existing data can be disaggregated for relevant groups of children and reported for different jurisdiction levels (nation, states, school districts, and individual schools).

Features of a System of Indicators

The information we reviewed led us to conclude that a set of indicators to monitor educational equity should focus both on valued student outcomes and on students' access to opportunities and resources needed to achieve those outcomes. The measures should reflect multiple dimensions of educational progress and well-being, including both academic achievement and engagement in schooling, since both contribute to students' likelihood of achieving valued outcomes, such as earning a high school diploma. The measures should be sufficiently precise and meaningful to identify important between-group disparities at single points in time and consistent enough to track them across years. To be useful to a variety of stakeholders, the indicator system should have the capacity to report results at multiple geographic and organizational scales, such as at the classroom, school, district, state, or national level. Critical for the system is the identification and definition of the population groups to be tracked. The system should allow for evaluation of disparities between salient, well-defined population groups.

We have already discussed the fact that students of color and students from financially disadvantaged families experience the K–12 public education process differently from their white and more affluent counterparts. In the indicator system we propose, we do not attempt to address the societal factors that underlie these differences, but we are aware that family and neighborhood factors play important roles in the educational resources available to students. The indicator system should include measures to evaluate and monitor availability of resources that bear on school learning, such as experienced teachers, safe schools, and strong curricula.

Reports of these indicators should be produced on a frequent and regular schedule so that stakeholders can anticipate their release, and they should be easily accessed and understood. The statistics that are reported should be based on data collected in a scientifically sound manner that can support the intended inferences. The indicator system will likely evolve as new kinds of data become available or new equity issues arise. Despite the system's overall focus on maintaining consistency to support the integrity of the trend information, its development should anticipate that changes will be needed.

Of course, an indicator system is merely information and cannot itself directly improve equity by altering policies and practices. Any project like ours rests on the familiar assumption that decision makers—officials, practitioners, voters—who have good, useful information will make better decisions. Admittedly, this is akin to an economist's assumption of rational behavior and may be just as problematic. However, a thorough exploration of how the behavioral sciences might bear on design and implementation of an indicator system is beyond the scope of this report.

> **CONCLUSION 2-1:** To be useful to policy makers, educators, and other stakeholders, two types of equity indicators are needed: (1) indicators that measure disparities in students' academic achievement and attainment outcomes and engagement in schooling; and (2) indicators that measure equitable access to resources and opportunities, including the structural aspects of school systems that may affect opportunity and exacerbate existing disparities in family and community contexts and contribute to unequal outcomes for students.

> **CONCLUSION 2-2:** To ensure that the pursuit of equity encompasses both the goals to which the nation aspires for its children and the mechanisms to attain those goals, a system of educational equity indicators should balance breadth of coverage with specificity to the appropriate stages of child development and to relevant groups facing disparity. It should also balance consistency across time and place with sensitivity to temporal and geographic context. The committee identified the following characteristics as crucial to striking this balance:
>
> 1. measure multiple dimensions of educational outcomes and opportunities, including changes over time;
> 2. focus on disparities between the population subgroups most salient for policy attention;
> 3. use measures that are comparable across time and place, and useful at several organizational scales (classrooms, schools, districts, states, nation);
> 4. use indicators and measures appropriate to grade level;
> 5. measure contextual and structural characteristics of or affecting the educational system, such as racial segregation and concentrated poverty;

6. produce frequent, readily understood, high-level summary statistics, in addition to more nuanced statistics;
7. be based on scientifically sound measures; and
8. incorporate mechanisms for continuous improvement based on research and other developments.

CONCLUSION 2-3: Existing data collection programs and related publications present a mixed picture with regard to their ability to support the committee's proposed set of K–12 educational equity indicators.

PROPOSED INDICATORS

Indicators of Disparities in Student Outcomes

The committee's charge calls for us to identify equity indicators for pre-K to grade 12 and then on to the transition to postsecondary education and work. We use a step ladder as a visual depiction of the stages of education from preschool to graduation and to emphasize important characteristics of the indicators we are proposing: see Figure 2-2.

- First, at each step, there are a multitude of outcomes that we could have selected. Since the point of an equity indicator system is to highlight a small set of statistics that are most useful for addressing disparities, we selected outcomes we judged to be milestones, critical for success at the next successive step. We based these judgments on our reviews of the literature, including both qualitative and quantitative empirical research, edited volumes that summarize this research, and thought pieces about research findings by experts in the field.
- Second, the outcomes reflect a progression: achievement at one step builds on achievement at the preceding step and in turn serves as the building block for the next step. What one learns in one step is cumulative and carries into next step.
- Third, the outcomes are predictive. The indicators at one stage are predictive (or strongly related to) the indicators at the next stage, and ultimately, they should be predictive of educational attainment.

At each step, we considered multiple dimensions of learning and achievement. Those dimensions included achievement in multiple subject areas (reading, math, science), multiple measures of achievement (e.g., performance in coursework, performance on standardized tests), and multiple aspects of learning, such as showing achievement progress and being engaged in what one is learning. We group factors related to student outcomes into three domains:

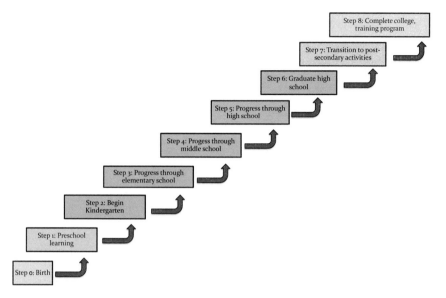

FIGURE 2-2 Steps in the educational process: Preschool to high school graduation and the transition to postsecondary pursuits.
NOTES: Steps shaded in purple, blue, and green are within the purview of this study; steps shaded in gray are outside the purview of this study.

- Domain A: Kindergarten Readiness
- Domain B: K–12 Learning and Engagement
- Domain C: Educational Attainment

Indicators of Disparities in Access to Educational Opportunities

Societal conditions such as income inequality and residential segregation intersect with the educational process in ways that have profound implications for efforts to reduce group disparities in educational progress, achievement, and attainment. These conditions lead to disparities in the resources that are available to support children's learning and development in their families, schools, and neighborhoods. Combined with the differential treatment of children and parents who are racial, ethnic, and linguistic minorities, these disparities contribute to between-group differences in educational outcomes. Between-group differences in education outcomes are important to understand and monitor because they reflect differences in the structure of educational opportunity—thus signaling disparities in the education system. The goals for education that we have described are easier to attain when students have access to well-resourced, high-quality preschool experiences and K–12 schools.

Many students do not have access to these important resources, and their access is strongly related to life circumstances beyond their control—where they live, their parents' education and income, their race or ethnicity, and the language spoken in their home. Schools cannot be expected to address the root causes of income inequality, residential segregation, or structural racism, but as long as these conditions exist, schools and school systems must grapple with their effects. Without meaningful actions on the part of schools, communities, and states, the education system will simply replicate societal disparities. Chapters 3 through 5 explore these factors in more detail. Chapter 3 discusses ways in which family, neighborhood, and societal factors shape a child or student's context in ways that affect school readiness and student outcomes generally. Several dimensions of these domains lack an adequate consensus or research base to fashion indicators and measures at this time. That is work for the future, contributing to continuous improvement of the indicator system.

We group factors related to opportunity and resources in the education system into four domains:

- Domain D: Extent of Racial, Ethnic, and Economic Segregation
- Domain E: Equitable Access to High-Quality Early Learning Programs
- Domain F: Equitable Access to High-Quality Curricula and Instruction
- Domain G: Equitable Access to Supportive School and Classroom Environments

Domains A through C are discussed in Chapter 4, and Domains D through G are discussed in Chapter 5.

3

Contextual Influences on Educational Outcomes: Families and Neighborhoods

Societal conditions, such as income inequality and residential segregation, and social institutions, such as families, intersect with the educational process in ways that have profound implications for efforts to reduce group disparities. These conditions lead to disparities in the resources that are available to support children's learning and development in their families, schools, and neighborhoods. Combined with the differential treatment of children and parents who are racial, ethnic, and linguistic minorities, these disparities contribute to between-group differences in important educational outcomes (Reardon, 2016; Ribar, 2015).

This societal/institutional framework is consistent with the emerging consensus among child development and neuroscience researchers. This "brain science" or psychosocial framework connects childhood and adolescent learning outcomes with the brain's responses to chronic stress and some forms of adversity; it includes a causal model based on neurochemistry, physiology, neuroanatomical plasticity, epigenetics, and related fields (Cantor et al., 2018; Osher et al., 2018). These two frameworks—the societal/institutional and the psychosocial/biomedical—together help in understanding the social determinants of learning. This framework is analogous to the social determinants of health, a familiar construct in medicine and public health.

With this framework in mind, this chapter discusses an interrelated set of contextual factors that affect academic engagement, progress, and attainment. Although these factors are largely beyond the control of the education system, we raise them because they strongly affect individual students, disproportionately affect certain groups of students, and merit an intentional societal response to improve educational equity.

FAMILY RESOURCES

Families have been called the smallest schools because of the vital role they play in children's education and development (Barton and Coley, 2007, p. 2):

> Indeed . . . the foundation established and nurtured at home goes a long way in ensuring student achievement in school as well as success in later life. The important educational role of parents, however, is often over-looked in our local, state and national discussions about raising student achievement and closing achievement gaps.

Some of the same characteristics that are important in schools also are important in families: financial resources, the number of children and adults present in the home (family structure), and the quality of the environment and relationships in that environment. Differences in these types of family resources and supports are associated with between-group differences in children's school-related outcomes.

Financial Resources

One of the most important resources families can have is sufficient income. Adequate income allows families to live in a safe, unpolluted neighborhood, have access to good schools, not worry about having enough money to pay the rent or having to move frequently—and possibly change schools—when rents increase, be able to afford preventative visits to the doctor, and be able to withstand emergencies. Yet the continually shrinking middle class and widening gaps between the highest and lowest income groups in the United States over the past several decades have left many families living in or very close to poverty and without this type of security. And the same structural inequalities that give rise to poverty can cause it to become chronic throughout a lifetime and to persist into the next generation.

The cumulative effects of socioeconomic disadvantage, absent effective interventions and supports, can have enduring effects on children. Research has shown that these effects are associated with several categories of adverse childhood experiences, some of which are listed in Box 3-1. Many are more common among low-income families and in high-poverty communities. For example, children who live in poverty experience more language delays, poorer nutrition, and more chronic illness than higher-

---

BOX 3-1
Some Adverse Childhood Experiences and
Sources of Chronic Stress Indicators

Sources of stress listed by the U.S. Department of Health and Human Services:[a]

- Physical abuse
- Sexual abuse
- Emotional abuse
- Physical neglect
- Emotional neglect
- Intimate partner violence
- Mother treated violently
- Substance misuse within household
- Household mental illness
- Parental separation or divorce

Some other sources of chronic stress:

- Homelessness and housing insecurity
- Food insecurity
- Neighborhood violence
- Juvenile/criminal justice system involvement, self or within family
- Immigration enforcement and risks

---

[a]Information from U.S. Substance Abuse and Mental Health Services Administration. Available: https://www.samhsa.gov/capt/practicing-effective-prevention/prevention-behavioral-health/adverse-childhood-experiences.
SOURCE: Pynoos et al. (2014).

---

income children (American Academy of Pediatrics, 2016; National Center for Children in Poverty, 2019). The challenges associated with poverty also influence children's readiness, attendance, engagement, and performance in school. In addition, children's health or the health of the adults in their households, lack of transportation, abuse, or neglect can affect school attendance (Balfanz and Byrnes, 2012a,b). Changes in family employment or living situations (including homelessness) may result in one or more moves during a school year, which can adversely affect academic progress (Duncan and Murnane, 2011). Students who arrive at school hungry, stressed, or suffering the consequences of maltreatment and other traumas may have difficulty concentrating. Their stress response systems may lead to behaviors that interfere with the learning process, which in turn may lead to disciplinary actions (Brooks-Gunn and Duncan, 1997; Duncan and

Magnuson, 2011). It is also difficult to study or complete homework assignments in some home environments, and these difficulties are compounded for the large number of children who come into contact with the child welfare, foster care, and juvenile justice systems each year.

When families live in precarious economic circumstances, parents also are more likely to struggle with physical and mental health challenges, such as depression and substance abuse (Adler et al., 1993; Brooks-Gunn and Duncan, 1997; Brooks-Gunn, Duncan, and Aber, 1997; Duncan and Brooks-Gunn, 1999; Duncan, Brooks-Gunn, and Klebanov, 1994). These challenges negatively affect the quality and stability of family relationships and make it more difficult for parents to provide a warm, responsive, supportive environment for their children. These challenges may be compounded for immigrant families, especially those who face potential deportation and family separation because one or more family members is undocumented. Chronic exposure to these kinds of conditions creates stressors that compromise children's neural and cognitive development and affect their health, well-being, and behavior (Blair and Raver, 2012; Cantor et al., 2018; National Scientific Council on the Developing Child, 2014; Osher et al., 2018). In the short term, exposure to the toxicity of adverse childhood experiences and chronic stress can interfere with children's self-regulation, executive function, learning, and memory (American Academy of Pediatrics, 2016). In the longer term, it can increase susceptibility to a variety of physical illnesses and mental health problems (National Scientific Council on the Developing Child and National Forum on Early Childhood Policy and Programs, 2011), which lie within the category of social determinants of health.

Often, these challenges accumulate, increasing the risk of school failure and other adverse life outcomes (Educational Testing Service, 2013). Indeed, achievement gaps on standardized tests between high- and low-income students in the United States have grown by 40 percent in a generation, and gaps have grown by the equivalent of 35 SAT points (on an 800-point SAT scale) (Reardon, 2011). These types of gaps ultimately lead to disparities in educational attainment, which further compounds inequities and can result in lack of economic mobility (Reardon, 2011).

Family Structure

The number of parents at home and the stability of parental relationships can also affect the types of family processes that promote children's educational success. The associations between family structure and children's outcomes arise for several reasons. First, having more adults in a household may translate into more time available for reading to children in early childhood or monitoring them during adolescence. Second, families with fewer adults in the household typically have fewer economic resources (Sigle-Rushton and McLanahan, 2002; Thomas and Sawhill, 2005). Third, family structure instability is associated with children's self-regulation and attention skills. These effects appear to be more pronounced if the father is absent during early childhood, and they may be more pronounced for boys than for girls (McLanahan, Tach, and Schneider, 2013). Although children who live apart from one of their parents are more likely to experience negative outcomes than children who live with both parents, family resources and processes are more important for children's outcomes than family structure itself.

Supports for Learning

Poverty and the absence of fathers can reduce investments of parental time and cognitive stimulation, both of which are crucial to children's development and learning. These investments are particularly important during the early years of life, when brain development and cognitive development are at their most plastic (Shonkoff, 2010). Although parenting and caregiving are beyond the scope of this report, we mention them here because of their importance for educational outcomes.

Differences in the quality and quantity of out-of-school learning supports are a major influence on group differences in children's learning and academic achievement (Bassok et al., 2016). In general, wealthier families are more likely to be able to afford materials, experiences, and services that support their children's development, such as books, computers, family educational activities, enrichment activities outside the home, and tutoring (Garrett, Ng'andu, and Ferron, 1994). In contrast, those investments may be less affordable for families with limited resources. Moreover, as mentioned, the living conditions of some families may not be as conducive to learning (e.g., low lighting, high noise levels, or limited space) (Dearing and Taylor, 2007; Evans, 2004).

Research has long shown that reading to young children is vital to helping children acquire important literacy skills (Leibowitz, 1977; National Research Council, 1998; Scarborough and Dobrich, 1994; Storch and Whitehurst, 2001). Children in higher-income families are read to more frequently than children in lower-income families. Children in two-parent families are also more likely to be read to than children in one-parent families (Federal Interagency Forum on Child and Family Statistics, 2012a, as cited in Educational Testing Service, 2013).

Time is often used as a proxy for investment in children's socioemotional and academic development, though the quality of time spent matters as much as, if not more than, the amount of time (Magnuson, 2018). Time-use data show significant variations by income level in the amount of time parents spend with their young children (Guryan, Hurst, and Kearney, 2008; Kalil, Ryan, and Corey, 2012). Self-report time diary data also have revealed meaningful differences in the developmental quality of time that mothers spend with their children by mother's level of education (Kalil, Ryan, and Corey, 2012; Ramey and Ramey, 2010).

NEIGHBORHOOD RESOURCES

The challenges described above are magnified in neighborhoods where many other families are poor and experience precarious circumstances. Moreover, such neighborhoods often lack institutional resources that can help protect children and parents from the effects of economic insecurity. As discussed in detail below, neighborhoods with a high concentration of low-income families tend to have higher crime rates and less access to healthy foods (i.e., "food deserts") and to lack basic resources for medical and dental care. These conditions take a toll on children.

Segregation and Economic Context

Neighborhood economic context has powerful, long-term effects on educational achievement and attainment (and earnings and family structure) (Chetty, Hendren, and Katz, 2016; Reardon, 2016; Schwartz, 2010). The historical legacy of racism, discrimination, and exclusion has disproportionately affected black children, who are more likely to experience precarious economic circumstances because they have grown up without familial wealth to rely on in times of crisis. They also are more likely to live in neighborhoods with other financially disadvantaged African-American families (Patillo, 2013; Reardon, Fox, and Townsend, 2015; Sharkey, 2010). The same is true for Latino children who experience the effects of discrimination and isolation as a consequence of language and cultural differences and immigration patterns.

Wealth inequality—in addition to housing and zoning policy reflecting discrimination—also has led to neighborhood and societal segregation (Owens, 2016; Reardon and Bischoff, 2011; Rothwell and Massey, 2010). While race-based neighborhood segregation has been slowly declining overall, socioeconomic segregation has steadily risen (Owens, Reardon, and Jencks, 2016; Reardon et al., 2018; Reardon and Bischoff, 2011). Socioeconomic segregation patterns shape children's residential contexts and the quality of the education, support services, and enrichment opportunities that are available to them (Putnam, 2015). Because children attend schools near their homes, school and neighborhood quality are linked. Schools in communities with abundant resources can draw on those resources in ways that schools in poorer communities cannot.

Residential segregation concentrates financially disadvantaged, black, and Hispanic children and families in high-poverty neighborhoods, which compounds the effects of poverty and magnifies societal inequalities. In segregated contexts, not only are children from lower-income families subject to the stresses and educational challenges associated with family-level poverty, but also lower-income families are much more likely to live in high-poverty neighborhoods and their children are much more likely to attend high-poverty schools (Patillo, 2013; Reardon, Fox, and Townsend, 2015; Sharkey, 2010). Conversely, children from higher-income families are more likely to live in high-income neighborhoods and attend more affluent schools.

To the extent that high-poverty contexts limit educational opportunities and high-income contexts expand opportunities, segregation will exacerbate inequalities in educational opportunity and outcomes. Indeed, racial differences in exposure to economically disadvantaged schoolmates are linked to achievement gaps, and these achievement gaps are larger in more segregated school systems (Condron et al., 2013; Reardon, 2016; Reardon, Kalgorides, and Shores, 2019).

Environmental Quality

An increasing body of research points to environmental conditions and hazards as threats to learning and development, and children in low-income neighborhoods are more likely to be exposed to these harmful conditions (Magnuson, 2018). Exposure to toxins, such as tobacco smoke, air pollutants, and lead, for example, can lead to a wide range of health and developmental problems, and exposure varies by poverty status. In 2010, 10 percent of children in families below the poverty level lived in homes where someone regularly smoked, compared with 3 percent of children in the most affluent families (Educational Testing Service, 2013). In terms of lead exposure, 21 percent of children aged 1-5 in families below the poverty level had 2.5 or more micrograms of lead per deciliter of blood, compared with 10 percent of children in families above the poverty level (Educational Testing Service, 2013).

Environmental and community conditions are now included as a factor that contributes to group differences in human capital accumulation (Dilworth-Bart and Moore, 2006; Magnuson, 2018). For example,

- In-utero residential proximity to environmental toxins in Superfund sites predicts children's later achievement (Persico, Figlio, and Roth, 2016).
- Lead abatement efforts are associated with improvements in children's school outcomes (Aizer et al., 2016; Sorenson et al., 2019).

- For school-aged children, exposure to community violence is associated with lower academic test scores (Margolin and Gordis, 2000; Sharkey, 2010; Sharkey et al., 2014).

Moreover, while the findings about academic achievement hold for K–12 students, the extent to which exposure to harmful environmental conditions explains group differences in kindergarten readiness is much less clear (Magnuson, 2018).

Children who experience any or all of these challenges in their families and neighborhoods need especially supportive schools. But as we discuss in Chapter 5, evidence suggests that the schools available in their neighborhoods tend to be less strong, at least on some dimensions—such as more novice teachers, fewer rigorous course offerings, and climates that do not support student learning—than the schools available to families with more means.

SAFETY, TRAUMA, AND CHRONIC STRESS

The accumulation of family and neighborhood risks detailed above is associated with increased occurrences of adverse childhood experiences and trauma, including child maltreatment and exposure to domestic and intimate partner violence in the families. Children in such contexts are also at increased risk of exposure to community violence, both as witnesses and as direct victims (Richters and Martinez, 2016). Children who live in high-poverty neighborhoods are also exposed to higher levels of homicide, which has been shown to be associated with lower test scores (Sharkey, 2010). Trauma may also be related to children's capacity for attention and impulse control, which are important for academic outcomes. For example, preschoolers who experienced recent local violence near their homes exhibited lower attention and impulse control than they had previously shown and performed less well on assessments of preacademic skills, with parents also showing more distress. This evidence highlights a potential pathway of effects of community violence on even very young children who may not themselves be aware of the incidents (Sharkey et al., 2012).

Research shows that exposure to violence has deleterious developmental effects in terms of the incidence of trauma symptoms, behavioral dysregulation,[1] impaired cognitive development, and their underlying

---

[1]Harmful behaviors that people use as they try to cope with stressful situations. These behaviors can include drinking alcohol to cope with problems, binge eating, extreme social reassurance seeking, and non-suicidal self-injuries (NSSI). These behaviors allow people to shift their attention away from unpleasant emotional states and toward bodily sensations, taste, and social support. Relief is only short-term and may trigger new behaviors, such as feelings of guilt, negative bodily states, and social problems (Jungmann et al., 2016).

neurobiological organization[2] (Margolin and Gordis, 2000). Understanding of the relationships between community and family violence and child well-being is incomplete but rapidly evolving. Efforts to learn more about children's physical and emotional responses to trauma can support the development and implementation of interventions.[3]

More broadly, as noted at the beginning of this chapter, the broader frameworks of societal/institutional and psychosocial/biomedical considerations encompass several aspects of context: see Box 3-1. It is important to note that the effect of adversity on a child or adolescent depends not only on individual resilience and natural variations in child development, but also on the child's opportunity for experiences, interventions, and supports that may mitigate or even undo the effects of adversity, both material and psychological. For example, within the psychosocial/biomedical framework, scientists point to neuroplasticity and the malleability of function and anatomy in response to experiences, relationships, and the general context. Brain malleability supports the notion that context need not be destiny; learning obstacles that are a result of context are not student deficits barring success, but student needs that can be met with appropriate opportunities.

The collection of potentially useful opportunities suggested by research is quite broad. Apart from obvious material examples, such as free lunches, the list includes caring adults, supportive peer relationships, mental health services, culture-informed pedagogy, and trauma-informed disciplinary and instructional strategies.

More research is needed to increase understanding of how various interventions or opportunities map onto individual student needs that are rooted in context. In addition, research and consensus-building are needed to create indicators and measures that could eventually be included in an equity indicator system.

For many student needs, screening and responses can best be provided outside of school settings, budgets, and systems. Educators and staff often lack adequate professional and fiscal resources. Therefore, an indicator system that encompasses all the domains of opportunity important for equity will need to monitor how well student success is supported by other child-serving agencies and nonprofit organizations.

---

[2]Neurobiology is the study of cells of the nervous system and the organization of these cells into functional circuits that process information and mediate behavior. See www.sciencedaily.com.

[3]The Substance Abuse and Mental Health Services Administration's National Child Traumatic Stress Initiative (NCTSI) represents the realization of congressional recognition of the serious impact on child mental health and well-being of the experience of traumatic events in childhood; it includes technical assistance and training in trauma-informed evidence-based modalities through the National Child Traumatic Stress Network: see https://www.samhsa.gov/child-trauma.

CONCLUSIONS

Out-of-school context matters for student outcomes. The cumulative effects of toxic stress in both environmental and social domains across childhood can affect school readiness, engagement, and achievement. Information about a student's context should be available to schools, limited by due attention to privacy and unreasonable intrusiveness.

CONCLUSION 3-1: The circumstances in which students live affect their academic engagement, progress, and attainment in important ways. If narrowing disparities in student outcomes is an imperative, schools cannot shirk the challenges arising from context. Neither can they confront these challenges on their own. Contextual factors that bear on learning range from food and housing insecurity to exposure to violence, unsafe neighborhoods, and adverse childhood experiences to exposure to environmental toxins. Children also differ in their individual responses to stress. Addressing student needs, in light of their life circumstances, requires a wide variety of resources. It is a responsibility that needs to be shared by schools, school systems, other agencies serving children and families, and nongovernmental community organizations.

In the chapters that follow, we focus on indicators of measurable student outcomes that may be influenced by family and neighborhood contextual factors, but we do not propose indicators of contextual factors themselves. Future efforts to target the root causes of disparities in student outcomes would require more direct measures of those family and neighborhood factors.

4

Indicators of Disparities in Student Outcomes

As described in Chapter 2, we propose indicators that fall into two categories: indicators of disparities in students' educational outcomes and indicators of disparities in students' access to educational resources and opportunities. This chapter addresses the first category.

We have chosen this set of indicators because they are measures of outcomes that we judged to be critically important milestones for success as students proceed from kindergarten through the postsecondary transition (see Figure 2-2 in Chapter 2). The proposed indicators are appropriate for different developmental stages (i.e., grade level): that is, they can measure the contours of (in)equity at different stages from pre-K through grade 12 and the transition to postsecondary activities. They also offer diagnostic capability: understanding when key inequities arise, narrow, or widen is useful for identifying targeted interventions.

Many of our proposed indicators have long formed the bedrock for measuring educational achievement and attainment. In addition, we propose indicators that offer opportunities to move beyond traditional measures of achievement and focus on different dimensions of key educational outcomes. Sound measures exist for many of the outcome indicators we propose, and several of them have been demonstrated to predict longer-term outcomes. Some of our indicators, however, are considered important in theory, but measurement is less well developed, and the research is not yet conclusive. In some cases, measures are available at the local or state level but not nationally so they cannot be compared across the country. Nevertheless, we propose these indicators envisioning that they will eventually be developed for use nationally. In the meantime, we encourage their use at the state, district, or school level.

This chapter has four sections: one for each of the three domains and a fourth one on the availability of data and reports for those domains. In Appendix C the committee provides illustrations of the data sources and methods that could be used to develop appropriate measures for our proposed indicators. For each domain, we briefly summarize the research base, identify the indicators we judge to be important, and discuss constructs to measure for these indicators.

Table 4-1 shows the committee's indicators for the three domains discussed in this chapter. For each domain (column 1), the table shows the indicators (column 2) and the constructs to measure them (column 3). This table provides a general framework for designing a system of equity indicators; we do not suggest the actual metrics or statistics for reporting on the indicators (e.g., percentages, averages). Rather, we discuss below the constructs to measure and ways to measure them. Some are ready for reporting on a national and subnational scale, disaggregated by population groups. Some are not. We discuss their status in the text.

DOMAIN A: KINDERGARTEN READINESS[1]

The first 5 years of life are a time of rapid learning and development that has profound and lasting effects. During this sensitive period, the developing brain is especially primed to create neural networks that support learning and development for years to come (National Research Council and Institute of Medicine, 2000). As discussed in Chapter 3, factors that interfere with the early building of the brain's architecture can cause developmental delays and associated challenges that can persist or are irreversible. Early childhood experiences set the stage for later academic success. From an equity perspective, monitoring kindergarten readiness is important because large between-group disparities become apparent well before children enter kindergarten (Halle et al., 2009; Howard and Sommers, 2015; Lee and Burkham, 2003) and can have lasting effects (Duncan and Magnuson, 2011; Huttenlocher et al., 2010).

Broadly speaking, kindergarten readiness is the set of foundational skills, behaviors, and knowledge that enable children to successfully transition into kindergarten and achieve academic success throughout the primary grades (Sabol and Pianta, 2017). The importance of early achievement in literacy and mathematics in forecasting later achievement in these areas has been well documented (Duncan et al., 2007; Duncan and Magnuson, 2011). Attention skills and cognitive self-regulation are also thought to be

---

[1]This section is drawn from Magnuson (2018).

TABLE 4-1 Proposed Indicators of Disparities in Student Outcomes

DOMAIN	INDICATORS	CONSTRUCTS TO MEASURE
A Kindergarten Readiness	1 Disparities in Academic Readiness	Reading/literacy skills Numeracy/math skills
	2 Disparities in Self-Regulation and Attention Skills	Self-regulation skills Attention skills
B K–12 Learning and Engagement	3 Disparities in Engagement in Schooling	Attendance/absenteeism Academic engagement
	4 Disparities in Performance in Coursework	Success in classes Accumulating credits (being on track to graduate) Grades, GPA
	5 Disparities in Performance on Tests	Achievement in reading, math, and science Learning growth in reading, math, and science achievement
C Educational Attainment	6 Disparities in On-Time Graduation	On-time graduation
	7 Disparities in Postsecondary Readiness	Enrollment in college, entry into the workforce, enlistment in the military

consequential to children's learning because they indicate the extent to which children are able to sit still, concentrate on tasks, persist at a task despite minor setbacks or frustrations, listen and follow directions, and work independently or, conversely, whether they are easily distracted, overactive, or forgetful. Studies have consistently found positive associations between measures of children's ability to control and sustain attention with academic gains in the preschool and early elementary school years (Brock et al., 2009; McClelland, Morrison, and Holmes, 2000; Raver et al., 2005).

Family income and education are strong correlates of kindergarten readiness. One explanation for this relationship is that parental education and income structure much of children's early lives in terms of their experiences both inside and outside of their homes. For example, a primary driver of early school readiness is how much time and cognitive stimulation children receive from their parents, other family members, and caregivers. The amount of time parents *spend* with their children (or *have available* to spend with their children) is related to family socioeconomic status (Guryan, Hurst, and Kearney, 2008; Kalil, Ryan, and Corey, 2012; Sayer, Bianchi, and Robinson, 2004). Children's participation in enrichment activities outside the home, such as arts and crafts, music, and physical activities (e.g., Gymboree, swimming lessons) is also predictive of early school readiness skills (Pre-Kindergarten Task Force, 2017; Pritzker, Bradach, and Kaufmann, 2015; Yoshikawa et al., 2013). Attending out-of-home day care or preschool increases the likelihood of exposure to these activities. These are investments that vary by family income level, and they seem to matter.

Based on our review of the research, we think that two aspects of school readiness should be included in an equity indicator system: (1) children's academic readiness in literacy and mathematics and (2) children's self-regulation and attention skills.

Indicator 1: Disparities in Academic Readiness

Achievement in the preschool and early-childhood years refers mainly to a set of reading- and math-related skills. Children who enter school with a basic knowledge of math and reading are more likely than their peers to experience later academic success (The Annie E. Casey Foundation, 2010; U.S. Department of Education, 2014). But children who are "behind at the starting gate" have a hard time catching up (Isaacs, 2012).

For early learners, reading-related skills encompass identification of upper- and lowercase letters as well as decoding skills such as beginning to associate sounds with letters at the beginning and end of words. Most early reading problems reflect poor decoding skills and low levels of phonological and phonemic awareness, such as a poor ability to break down words into component sounds. As children progress through childhood, reading skills include recognizing words by sight, understanding words in context, and making literal inferences from passages.

Concrete math skills begin with the ability to recognize numbers and shapes and to compare relative sizes. Counting and sequencing skills are followed by the ability to perform addition and subtraction tasks, as well as multiplication and division tasks. Understanding numerical properties such as proportions, fractions, integers, and decimals also develops, as do measurement skills and an understanding of geometry. These pre-academic and academic skills develop as a result of learning opportunities embedded in everyday activities and specific instruction, which is especially important for code-related reading skills and computational mathematical skills.

Between-group differences in measures of academic readiness (recognize all letters, count to 20, write name, read words in a book) vary according to two key parental characteristics, education attainment and home language:

- Parental education attainment: Among 3- to 6-year-old children (not yet enrolled in kindergarten) whose parents had not completed high school, 15 percent could recognize all letters of the alphabet, 38 percent could count to 20 or more, 37 percent could write their name, and 13 percent could read words in a book. These figures are between 46 and 142 percent lower than those for children whose parents had completed some college or a vocational program and between 66 and 224 percent lower than those for children whose parents obtained a bachelor's degree.[2]
- Home language: Children with at least one parent who spoke English were more likely to demonstrate school readiness than those with two parents who did not speak English. In terms of letter recognition, 24 percent of children whose parents did not speak English could recognize all 26 letters, compared with 29 percent of children with one parent that spoke English and 41 percent of children with two English-speaking parents. For counting, 48 percent of children whose parents did not speak English could count to 20 or higher, compared with 52 percent of children with one English-speaking parent and 71 percent with two English-speaking parents.[3]

Indicator 2: Disparities in Self-Regulation and Attention Skills

Cognitive self-regulation is the "processes by which the human psyche exercises control over its functions, states, and inner processes" (Baumeister and Vohs, 2004, p. 1). It involves the ability to evaluate the steps and actions required to meet a desired goal and to control behavior deliberately in order to reach that goal. It is conceived of as a broad construct that includes multiple overlapping subcomponents, such as executive function, planning, sustaining attention, task persistence, and the inhibition of impulsive responses. Among experts in child development, there is no consensus on an exact definition, but nearly all agree that being able to sit in a classroom and pay attention is essential to school learning.

---

[2]See https://nces.ed.gov/ecls/index.asp.
[3]See https://nces.ed.gov/ecls/index.asp.

Cognitive self-regulation is often measured by checklists and observation protocols, completed by parents and teachers. Assessments may indicate the extent to which children are able to sit still, concentrate on tasks, persist at a task despite minor setbacks or frustrations, listen and follow directions, and work independently or, conversely, whether they are easily distracted, overactive, or forgetful.

Studies have consistently found positive associations between measures of children's ability to control and sustain attention and academic gains in the preschool and early elementary school years (Brock et al., 2009; McClelland, Morrison, and Holmes, 2000; Raver et al., 2005). However, there are considerable unresolved questions about whether a particular dimension of cognitive self-regulation (or behavioral manifestation of these skills) matters more than other dimensions or even if the associations can be interpreted as causal (Fuhs et al., 2014; Willoughby, Kupersmidt, and Voegler-Lee, 2012).

Differences in self-regulation and attention skills have been reported for students grouped by the socioeconomic status (SES) of their households. For example, research conducted as part of the Early Childhood Longitudinal Study, Kindergarten Class of 2010–11 (ECLS-K:2011)[4] documented readiness in terms of seven approaches-to-learning behaviors: paying attention in class, persisting in completing tasks, showing eagerness to learn new things, working independently, adapting easily to changes in routine, keeping belongings organized, and following classroom rules. Analyses revealed that students from lower-SES households tended to have lower approaches-to-learning scores than students from middle-SES and high-SES households.[5]

Proposed Measures for Indicators 1 and 2

Assessing readiness skills of incoming kindergartners is now a common practice in this country. Many states and districts have adopted assessments of the early literacy, numeracy, and socioemotional skills they deem import-

---

[4]See https://nces.ed.gov/ecls/index.asp. To date, the study includes three cohorts. As described on the website of the National Center for Education Statistics: "The birth cohort of the ECLS-B is a sample of children born in 2001 and followed from birth through kindergarten entry. The kindergarten class of 1998-99 cohort is a sample of children followed from kindergarten through the eighth grade. The kindergarten class of 2010-11 cohort is following a sample of children from kindergarten through the fifth grade. The ECLS program provides national data on children's status at birth and at various points thereafter; children's transitions to nonparental care, early education programs, and school; and children's experiences and growth through the eighth grade. The ECLS program also provides data to analyze the relationships among a wide range of family, school, community, and individual variables with children's development, early learning, and performance in school" (https://nces.ed.gov/ecls/kindergarten2011.asp).

[5]See https://nces.ed.gov/fastfacts/display.asp?id=680.

ant and that are aligned with their educational programs. There is wide variation in the assessments used and the measures that are reported, however, making it unlikely that results could be aggregated in a way that would support valid comparisons at a national or state level. Starting school ready to learn is essential for future success, but as documented above, something that varies widely as a consequence of family circumstances. Nevertheless, we argue that it is critical to monitor disparities in school readiness, even if only at a local level.

Over the years, there have been discussions about expanding NAEP's grade coverage to kindergarten (see, e.g., National Research Council, 2012). We see value in developing a version of NAEP that would be appropriate to measure kindergarten readiness. Having a set of standardized and uniform assessments of children's early skills, administered and reported on a regular basis, may be the only way that disparities in kindergarten readiness will receive the attention they deserve. This work could be informed by the assessments developed for the ECLS-K (1998 and 2010). We encourage efforts to develop such assessments, although such a program should be conscious of the toll testing can take on young children. We also note ongoing work to develop a national-level kindergarten readiness indicator that draws upon the National Survey of Children's Health and that can be disaggregated by race/ethnicity, income, parental education, and other relevant groups.[6]

DOMAIN B: K–12 LEARNING AND ENGAGEMENT

What students learn and how they perform in school positions them for future success. Course taking, course grades, and scores on tests are ways to measure students' achievement and progress in school, and many kinds of measures are available. The extent to which a student is interested in school and participates in his or her learning, often referred to as "engagement," also can have strong effects on academic performance and school completion. Learning and succeeding in school requires active engagement. Its opposite—disengagement—is associated with school failure and dropping out. Engagement (or disengagement) is important to monitor because most students do not just suddenly drop out: rather, they tend to go through a gradual process, sometimes called "stopping out," during which absences and tardiness increase, grades decline, and interest in school wanes. Dropping out is the final stage of disengagement.

Although engagement in learning is important for all students, the consequences of disengagement tend to be more serious for students from disadvantaged backgrounds. When students from advantaged backgrounds experience disengagement, their grades and school attendance may decline

---

[6]See https://www.childtrends.org/project/kindergarten-readiness-national-outcome-measure.

and they may learn less than they might have, but most eventually graduate and move on to other opportunities (National Research Council and Institute of Medicine, 2004). In effect, they often get a second chance. In contrast, the consequences of disengagement for middle and high school youth from disadvantaged backgrounds are severe; they are less likely to graduate from high school than their peers and face limited employment prospects, increasing their risk for poverty, poor health, and involvement in the criminal justice system (National Research Council and Institute of Medicine, 2004).

Indicator 3: Disparities in Engagement in Schooling

Recent work on engagement in learning consistently describes it as a multidimensional construct reflecting behavioral, emotional, and cognitive components (Appleton, Christenson, and Furlong, 2008; Fredricks, Blumenfeld, and Paris, 2004). Engagement is sometimes used interchangeably with motivation, but the two constructs have important differences. Broadly speaking, motivation is what drives a given behavior, and engagement is the outward manifestation of motivation (Fredericks and McColskey, 2012, p. 764).

Engagement in schoolwork involves both behaviors (e.g., persistence, effort, attention) and emotions (e.g., enthusiasm, interest, pride in success; Connell and Wellborn, 1991; Newmann, Wehlage, and Lamborn, 1992; Skinner and Belmont, 1993; Smerdon, 1999; Turner, Thorpe, and Meyer, 1998). Behavioral engagement refers to participation in the schooling process and includes involvement in academic, social, and extracurricular activities. It is sometimes defined in terms of positive conduct, such as attending school, completing assigned work, and adhering to classroom norms, as well as the absence of negative conduct, such as skipping school and disruptive behaviors (Connell and Wellborn, 1991; Finn, 1989; Finn, Pannozzo, and Voelkl, 1995; Finn and Rock, 1997; Fredericks and McColskey, 2012; Wang and Eccles, 2013). Behavioral engagement is often measured directly through self-reports of participation in extracurricular activities or teacher ratings of students' behavior in class, but it is also measured indirectly with administratively captured data, such as attendance, homework completion, tardiness, and suspensions (Fredericks and McColskey, 2012). Behavioral engagement is observable by others, making it a useful signal of how students are experiencing their academic environment.

Emotional engagement denotes positive affective school relationships with teachers, classmates, academic subjects, and the school as well as a sense of belonging (Dawes and Larson, 2011; Immordino-Yang, 2016). Emotional engagement is hard to observe directly, and many studies of emotional engagement use proxy measures, often based on behavioral engagement. However, there are instruments that capture students' self-reports of emotional engagement and mindsets about their work, as well as measures for teachers and parents that indicate their perceptions of students' emotional engagement (National Research Council and Institute of Medicine, 2004).

Cognitive engagement refers to student's level of investment in learning and the degree to which they are putting in effort to process material. Fredericks and McColskey (2012) characterize it as "being thoughtful, strategic, and willing to exert the necessary effort for comprehension of complex ideas or mastery of difficult skills (Corno and Mandinach, 1983; Fredricks, Blumenfeld, and Paris, 2004; Meece, Blumenfeld, and Hoyle, 1988). As with emotional engagement, cognitive engagement is not directly observable. A number of self-report survey measures capture different aspects of cognitive engagement. The University of Chicago Consortium on School Research has a measure of cognitive engagement that is reliable as either a student-level or school-level measure (Levenstein, 2014) and that is correlated with students' rates of passing courses and grade point averages (GPAs), separate from their test scores and background characteristics (Allensworth and Easton, 2007).

While engagement is experienced by students, it is a reflection of the interaction of students with the classrooms and school contexts in which they are functioning. For example, the same student will show different levels of engagement in different classes, with different teachers, peers, and subjects. This contextual dependence makes engagement difficult to measure and track over time and thus challenging to include in an indicator system.

Academic Engagement

Students' level of engagement can be evaluated through student self-report surveys, checklists completed by teachers, observations, and interviews. Instruments that measure engagement typically ask students to report on their attention, attendance, time on homework, preparation for class, class participation, concentration, participation in school-based activities, effort, adherence to classroom rules, and risk behaviors. A variety of surveys and rating scales are available, but none is in wide enough use to support its inclusion in a national indicator system. It can be included in local equity systems, however.

Attendance

To benefit from instruction, students must be at school. As discussed below, the positive relationship between instruction time and learning is well documented. When students fail to go to school, when absenteeism becomes chronic (e.g., missing 10% or more of enrolled school days; missing more than 15 school days), it can severely interfere with learning. Chronic absence is a powerful predictor of achievement because it means students have missed a substantial portion of instructional time over the course of the school year.

Chronic absenteeism affects students at all grades. Much of the early research on attendance focused on children in elementary, middle, or high school. However, more recent studies document that chronic absenteeism is a significant problem even among younger students, with 11 percent of kindergarteners nationwide chronically absent (Romero and Lee, 2007, cited in Allensworth and Easton, 2007).

Chronic absenteeism negatively affects student outcomes, and the impact is often greater for students in disadvantaged circumstances than for other students (Ehrlich, Gwynne, and Allensworth, 2018; Gottfried, 2014). This is true at all grade levels, from Head Start pre-K programs to high school (Allensworth et al., 2014; Ansari and Purtell, 2017; Aucejo and Romano, 2016; Gershenson, Holt, and Papageorge, 2016; Gottfried, 2009, 2010, 2014; Neild and Balfanz, 2006; Ready, 2017; Smerillo et al., 2018; Wang and Benner, 2014).

The Civil Rights Data Collection reported chronic absenteeism rates by population group for the 2015-2016 school year:

- Black students were 40 percent more likely to be chronically absent (missing at least 15 days during the school year) than white students: the rates were 20.5 percent for blacks and 14.5 percent for whites.[7]
- The same was true for Latino students, with a chronic absenteeism rate of 20 percent.[8]
- English-language learners (13.7%) were less likely to be chronically absent than their English-proficient counterparts (16.2%),[9] although this varied by race and ethnicity. For example, Latino English-learners were more likely to be chronically absent than English-learner non-Latinos.
- Students with disabilities were more likely to be chronically absent (22.5%) than their nondisabled counterparts (14.9%).

---

[7]Also see https://www2.ed.gov/datastory/chronicabsenteeism.html and https://datacenter.kidscount.org/data/tables/10125-chronic-absenteeism-by-race-and-other-category.

[8]Data from Chang, Bauer, and Byrnes (2018).

[9]A study by the UChicago Consortium on School Research found that much of the difference in 9th-grade grades and pass rates between students with disabilities and other students was explained by differences in attendance rates. The study also shows large differences based on disability type (Gwynne et al., 2009).

The rate of chronic absenteeism is higher in schools with high concentrations of students whose families are financially disadvantaged than in schools with lower concentrations of financially disadvantaged students. The U.S. Government Accountability Office (2018, p. 87) reported the following:

- For low-poverty schools (rate of 25% students in poverty or lower), the average rate of chronic absenteeism was 9.4 percent.
- For schools with a poverty rate between 50 and 75 percent, the average rate of chronic absenteeism was 13.1 percent.
- For high-poverty schools (rate of 75% or higher), the average rate of chronic absenteeism was 15.5 percent.

Proposed Measures for Indicator 3

Measures of chronic absenteeism can be included in a national indicator system; measures of academic engagement are likely available at the local level.

Indicator 4: Disparities in Performance in Coursework

While coming to school and behaving appropriately are necessary for learning, they are not sufficient. Students must learn what is being taught and demonstrate their learning by doing well in their courses. Course performance is important because it reflects the persistence, conscientiousness, and motivation needed to go to school each day, do the work, complete assignments, and turn them in on time. Research has shown that students' day-to-day performance in the classes they take, as represented by their course grades, is a strong predictor of on-time high school graduation, and, likewise, poor performance—especially course failure—is a warning sign of dropping out.[10]

High school course grades are also highly predictive of college grades and college graduation, as discussed below. There are considerable differences by race, gender, income, and disability status in students' grades and rates of passing classes (Jacob, 2002). There is also an intersection between the courses students take and their performance in those courses, so that

---

[10]Studies that found that grades or course passing in grades prior to 10th grade are predictive of high school graduation, or that found milestones strongly associated with graduation, such as passing an exit exam or being on track to graduate in 11th grade, include Allensworth and Easton (2005, 2007); Allensworth et al. (2014); Balfanz and Byrnes (2006); Balfanz, Byrnes, and Fox (2015); Balfanz, Herzog, and MacIver (2007); Baltimore Education Research Consortium (2011); Bowers (2010); Bowers et al. (2013); Hartman et al. (2011); Kieffer and Marinell (2012); Kurlaender, Reardon, and Jackson (2008); Neild and Balfanz (2006); Norbury et al. (2012); Stuit et al. (2016); Zau and Betts (2008).

some students obtain credits in a more varied curriculum than others or have differential likelihood of taking and passing courses in particular areas, particularly high-level science and math.

Societal inequities influence students' course selection and their ability to engage fully and successfully in their coursework and earn high grades. At the same time, school structures and classroom practices moderate the influence of societal factors on students' engagement and the likelihood that they will pass their classes. For example, students' attitudes toward science have been shown to be more positive in classrooms with strong teacher support, order and organization, and teacher innovation (Fouts and Myers, 1992). Student learning is stronger in classes with strong classroom control and challenging instruction (Bill and Melinda Gates Foundation, 2010; National Council of Teachers of Mathematics, 2014; Stein and Lane, 1996), while students' engagement, work effort, and course grades are stronger in classes where teachers support students through clear instruction, monitoring, and assistance (Allensworth et al., 2014). Students' exposure to high-quality instruction varies in relation to their backgrounds, with students from low-income backgrounds and minority students more likely to experience poorer-quality instructional environments (Ferguson, Stegge, and Damhuis, 1991; Hanushek, Kain, and Rivkin, 2004; King, Shumow, and Lietz, 2001; Oakes, 1990). Similarly, pass rates, credit accumulation, and GPAs show considerable between-group differences that reflect students' gender, race, ethnicity, and disability status.

Some states and districts have adopted single or composite indicators of being "on-track-to-graduate" or "at risk of school failure" that incorporate information about passing and failing courses for their accountability systems, or early warning indicators in their dropout prevention systems. For example, 11 states included a 9th grade on-track indicator in their Every Student Succeeds Act (ESSA) accountability system in 2018 (see Achieve, Inc., 2018). The indicator was defined on the basis of credit accumulation, sometimes in combination with the types of courses students were taking (e.g., accumulating credits in courses needed for graduation), and course performance.

Success in Classes

Students need to pass courses and accumulate credits in order to graduate high school. Beyond having a direct influence on credit accumulation, course failures can affect students' mindsets about whether they can succeed and belong in an academic environment; this influences their subsequent motivation and effort in school (Farrington et al., 2012). Grades and pass rates are less well documented across schools and districts than test scores, making it difficult to see the existing disparities.

Accumulating Credits across the Curriculum

Students need to pass their classes to make progress toward graduating high school, but there are also differences in the types of credits they are accumulating. Because of these differences, some states have chosen to incorporate the types of credits students accumulate into their metric for being "on track." There are a number of ways in which there are disparities in the types of courses students take, including whether students are taking college preparatory courses or a more basic curriculum; advanced placement courses that can count for college credit; courses in science, technology, engineering, and math (STEM); world language; and courses in the arts. These courses have implications for students' development of broad competencies and career options.

For example, several studies document a correlation between taking math and science classes during high school, particularly advanced classes in these subjects, and choosing a major in STEM and persisting with this major until graduation (Elliott et al., 1996; Maple and Stage, 1991; Riegle-Crumb and Humphries, 2012; Trusty, 2002; Wang, 2013; Ware and Lee, 1988). Yet there are large differences in the participation of students in STEM based on race, ethnicity, and income (Carnevale, Smith, and Melton, 2011; National Research Council, 2010; Runningen, 2014).

The kinds of classes that schools offer and the ways in which students are assigned to courses affect whether or not students take a broad curriculum and higher-level courses. For example, low-income students are often assigned to low-level math and science courses as early as the 9th grade and take the minimum STEM courses required for graduation, never developing higher-level math and science skills (Gamoran et al., 1997; Riegle-Crumb, 2006). Districts in areas that have a lower tax base to support schools often reduce offerings in the arts when they fall short on revenue or require families to cover the costs of arts programs, which limits the ability of students in low-income families to participate.

Grades and GPAs

Students' grades and GPAs are the strongest predictors of whether students will graduate from high school, showing more predictive power than test scores, attendance, pass rates, demographic factors, or students' families' income. High school GPAs also have the strongest evidence base as an indicator of readiness for college enrollment, college grades, persistence, and college completion (Allensworth and Clark, 2018; Bowen, Chingos, and McPherson, 2009; Camara and Echternacht, 2000; Geiser and Santelices, 2007; Geiser and Studley, 2002; Roderick, Nagaoka, and Allensworth, 2006). High school GPA not only is highly predictive of college outcomes, but also shows a generally consistent relationship with college outcomes across different high schools and colleges and is a strong predictor when comparing students with similar backgrounds and in the same high schools and colleges (Allensworth and Clark, 2018; Bowen, Chingos, and McPherson, 2009).

Several studies suggest that the threshold of a 3.0 high school GPA is the point at which students' probability of graduating college becomes greater than 50 percent, among those students who enroll in a 4-year college (Bowen, Chingos, and McPherson, 2009; Roderick, Nagaoka, and Allensworth, 2006). Differences in grades by gender, race, and disability status in middle school and high school are echoed in differences in high school and college completion many years later. Diprete and Buchmann (2013, p. 93), for example, show that 8th grade GPAs are highly predictive of college completion and explain the wide disparities in college completion by gender.

Proposed Measures for Indicator 4

Measures for Indicator 4 should include success in classes, accumulation of credits toward graduation, and GPAs or grades. These measures are not yet available at a national level, but they may be available for some districts and states. In particular, they may be available in the form of the on-track indicators many states and districts are developing.

Indicator 5: Disparities in Performance on Tests

Standardized student achievement tests have been the central feature of state accountability and reporting systems for decades; National Assessment of Educational Progress (NAEP) has been the primary way that the public tracks educational progress on a national level. Standardized test scores have several features that make them useful for monitoring students' educational attainment: they can provide a common metric across jurisdictions, they measure achievement in subjects that are core to most schools' missions, and they summarize information about student performance in a concise way. In addition, performing well on tests can open up opportunities for students, including admission to postsecondary institutions and access to scholarships. Thus, between-group differences in test scores are of concern not only because they may represent underlying inequities in attainment, but also because they could illuminate opportunity gaps. However, standardized test scores have limitations; as noted above, research shows that GPAs are a stronger predictor of later outcomes than test scores. Moreover, some tests may promote inappropriate inferences because of such factors as bias or lack of alignment with the full range of valued schooling outcomes. The committee acknowledges that test scores provide valuable information about students' skills and about gaps, but they should be supplemented with other information.

Between-group differences in performance on achievement tests are reported by NAEP. They can be measured at the national, state, and district level for some large urban districts. In 2017, when NAEP was last administered, substantial gaps in mathematics and reading achievement were evident across racial, ethnic, and SES subgroups, as well as for English learners and students with disabilities. For instance:

- White and Asian students' average scale scores in 4th grade reading (232 and 239, respectively)[11] were higher than the average scores of black (206) and Hispanic (209) students. Gaps in mathematics, and in both subjects at other grade levels, followed a similar pattern.
- Using eligibility for free or reduced-price meals as an indicator of socioeconomic status,[12] NAEP scores indicate that economically advantaged students consistently outperform disadvantaged students. On the 8th-grade mathematics assessment, for example, students who were eligible for free or reduced-price meals received an average score of 267, compared with 296 for students who were not eligible. The corresponding percentages of students scoring at the proficient or advanced levels were 18 percent and 48 percent for these groups.
- Performance gaps between students with and without disabilities are particularly large. On the 8th-grade reading assessment, average scale scores for these groups were 232 and 271, respectively.[13] NAEP allows students with disabilities to test with accommodations that are intended to enable them to access the test and to prevent their disability from threatening the validity of scores.
- Students who are classified as English learners typically receive lower test scores than native speakers of English.[14] These two groups received average scale scores of 246 and 285, respectively, on the 8th-grade math assessment, and scores of 226 and 269 on the 8th-grade reading assessment.

Although documenting disparities in performance on tests at a single point in time provides valuable information about between-group differences, these status-based measures do not provide any information about

---

[11]See https://www.nationsreportcard.gov/ndecore/xplore/NDE.

[12]See https://www.nationsreportcard.gov/ndecore/xplore/NDE.

[13]See https://www.nationsreportcard.gov/ndecore/xplore/NDE.

[14]See https://www.nationsreportcard.gov/ndecore/xplore/NDE.

the degree of progress students make as they move through the education system. Measures of achievement growth can support inferences about disparities in student progress: consequently, they can help users of the indicator system understand whether performance gaps that are observed when students start school increase or diminish as students are exposed to the programs and services that schools offer. This information can also shed light on the developmental phases during which students in different groups experience growth, stagnation, or regression in their skills.

It is important to point out that "growth" in this context does not simply refer to a change in scores, but instead encompasses a range of models that shed light on how performance of a student or group of students at one time compares to performance at a previous time (Castellano and Ho, 2013; Data Quality Campaign, 2019).[15] Growth models can be calculated on the basis of changes in individual students' performance or on changes in performance of an aggregate, such as a school.[16]

Most states have incorporated growth measures as part of their accountability indicators for ESSA (Data Quality Campaign, 2019). Published data on group differences in growth are less widely available than data on differences in average test scores. An example of such data is provided in a 2012 study by ACT that examined value-added achievement scores as well as simple change scores between 8th and 12th grades for students taking the ACT and EXPLORE assessments. Both approaches to measuring growth indicated that black and Hispanic students experienced less growth than white and Asian students.

---

[15]See https://scholar.harvard.edu/files/andrewho/files/a_pracitioners_guide_to_growth_models.pdf.

[16]The decision about what type of growth model to use needs to be informed by the features of the test and the inferences that users intend to make. For example, some tests use a vertical scale that allows for the calculation of simple change scores that can be loosely interpreted as indicators of the magnitude of student achievement growth, but tests that do not have a vertical scale cannot support this type of metric. Many growth models that are used for large-scale reporting, such as in-state accountability systems, rely on complex multivariate models (e.g., value-added models) or on conditional status models (e.g., student growth percentiles). Both models attempt to provide an indicator of how students or schools are performing relative to where they started out, but they do not measure "growth" directly.

A growth model that would be suitable for use in an equity indicator system would need to accommodate a standardized testing landscape that varies by state and that for the most part does not include tests with vertical scales. Opportunities to measure year-to-year growth for specific subjects and age groups will vary across states and districts. Some states, for instance, administer science or social studies tests in consecutive grades to elementary and middle school students; others test in these subjects only in a small subset of grade levels.

Proposed Measures for Indicator 5

After considering the relative benefits of the different types of achievement metrics discussed above, the committee concluded that indicators of equity in test performance should include two metrics: (1) average scale scores on measures of achievement in grades 4, 8, and one high school grade for math, English language arts, and the sciences; and (2) measures of achievement growth, using student-level data if possible, in math and English language arts for students in grades 4 through 8. Together, these two types of measures would provide valuable information about students at a given time, and how their performance changes as they move through the education system. Both are important from an equity perspective. The proposed metrics do not cover every subject and grade level that might be of interest, but they are likely to be the most feasible given the state and national testing regimes that are in place, and they provide evidence of student performance at several key milestones.

DOMAIN C: EDUCATIONAL ATTAINMENT

Education is a critically important way for individuals to pursue their goals in life. With a high-quality education, individuals are better prepared to choose a path toward productive and purposeful adulthood, whether that path is 2- or 4-year college, the labor force, or the armed forces. Collectively, a high-quality education for all means better informed and more productive citizens, which has consequences for the overall economic, physical, and civic health and well-being of the country. Research consistently shows between-group differences in educational attainment related to people's race, ethnicity, and gender (see Table 4-2).

Decades of research and data unequivocally show that the more education people have, the better off they are financially, emotionally, and physically. Higher levels of educational attainment are associated with higher salaries, job satisfaction, job security, and job benefits, all of which provide individuals with more economic freedom of choice. Educational attainment also is associated with a range of positive health behaviors and outcomes. At lower levels of educational attainment, these benefits decrease. For individuals who do not earn a high school diploma, the economic, social, and health consequences are especially severe (Currie, 2009; Cutler and Lleras-Muney, 2006; Day and Newburger, 2002; Harlow, 2003; Heckman and LaFontaine, 2010; Sum et al., 2009; U.S. Department of Labor, 2013a; Wong et al., 2002).

If educational attainment opens the door to a better life, then opportunities for educational attainment must be equally available to all students. Given the lifelong benefits that accrue with increasing levels of education,

TABLE 4-2 Educational Attainment of the Population Aged 25 and Older by Race, Ethnicity, and Nativity Status, 2015: in Percent

	High School Graduation or More	Some College or More	Associate's Degree or More	Bachelor's Degree or More	Advanced Degree or More
Population 25 and Older	88.4	58.9	42.3	32.5	12.0
Race and Ethnicity					
White Alone	88.8	59.2	42.8	32.8	12.1
Non-Hispanic White Alone	93.3	63.8	46.9	36.2	13.5
Black Alone	87.0	52.9	32.4	22.5	8.2
Asian Alone	89.1	70.0	60.4	53.9	21.4
Hispanic (of Any Race)	66.7	36.8	22.7	15.5	4.7
Nativity Status					
Native Born	91.8	61.3	43.3	32.7	11.9
Foreign Born	72.0	47.6	37.6	31.4	12.5

SOURCE: Ryan and Bauman (2016, p. 2); data from the 2015 Current Population Survey.

this committee's aspiration is for all students to have the opportunity to earn a 2- or 4-year college degree. This goal includes high-school graduation, readiness for postsecondary education, and postsecondary matriculation and completion. Because postsecondary persistence and completion are beyond the scope of this report, our indicators are focused on readiness for the transition to 2- or 4-year postsecondary education. However, our intent is not to diminish the importance of completing college, nor to ignore the sizable between-group differences in college enrollment, persistence, and completion rates (The Pell Institute for the Study of Equal Opportunity in Higher Education, 2017). Instead, our focus is on what the K–12 education system can monitor and act on to increase equity in postsecondary readiness and matriculation, which may help to improve equity in educational attainment.

Indicator 6: Disparities in On-Time Graduation

Graduating from high school on time and with a diploma remains one of the most critical educational objectives. It also paves the way to a multitude of better life outcomes, including the likelihood of attending college (Belfield and Levin, 2007; Oreopoulos and Salvanes, 2011; Rumberger, 2011).

The economic costs of dropping out of high school are steep and have become worse over the past 30 years, with dropouts earning dramatically less income and being more likely to experience unemployment than high school graduates (Day and Newburger, 2002; Heckman and LaFontaine, 2007; Sum et al., 2009). Dropouts also have a much higher risk than high school graduates of incarceration (Harlow, 2003; Sum et al., 2009), and they are more likely to engage in a range of behaviors that endanger their health (Centers for Disease Control and Prevention, 2008; Cutler and Lleras-Muney, 2006; Manlove, 1998; McLanahan, 2009; Pleis and Lucas, 2009). As a result, dropouts typically live shorter, less healthy lives than high school graduates (Currie, 2009; Cutler and Lleras-Muney, 2006; Wong et al., 2002).

The standard for measuring graduation rates is the adjusted cohort graduation rate (ACGR). The ACGR represents the percentage of entering 9th-grade students who earn a regular diploma within 4 years. The ACGR is based on individual-level, longitudinal student records that produce an authentic measure of the percentage of students who graduate. Because the records are based on individual-level data, ACGRs can be disaggregated to enable comparisons among different groups of students across schools, districts, and states and across time. As a result, they are very useful for research, policy, and accountability decisions (National Research Council, 2011, p. 112).

Though smaller than in the past, disparities in high school graduation rates by racial, ethnic, and other demographic factors remain substantial. For 2015–2016, McFarland et al. (2018, p. 130, Fig. 2) report the following:

- For race and ethnicity, the rate was highest for Asian/Pacific Islander students (91%) and white students (88%) and lowest for Hispanic students (79%), black students (76%), and American Indian/Alaska Native students (72%).
- The rate was 67 percent for students with limited English proficiency.
- The rate was 66 percent for students with disabilities.
- The rate was 78 percent for economically disadvantaged students.

Proposed Measures for Indicator 6

The adjusted cohort graduation rate can be used to measure on-time graduation.

Indicator 7: Disparities in Postsecondary Readiness

Of students who completed high school in 2016, 70 percent enrolled in college immediately following graduation, with 24 percent enrolling in 2-year colleges and 46 percent enrolling in 4-year colleges (McFarland et al., 2018, pp. 150-153). Looking at group differences, these data show:

- Asians have the highest college enrollment rate, at 87 percent.
- For other racial and ethnic groups, 71 percent of white and Latino high school graduates and 56 percent of black graduates enrolled immediately in college.
- Students from high-income families (those in the the top 20%) were much more likely to enroll in college than low-income students (those in the bottom 20%), 83 percent and 67 percent, respectively, while middle-income students were even slightly less likely to enroll than low-income students, 64 percent.

Despite widespread agreement about the need for the K–12 education system to focus on college readiness, there is no consensus on an evidence-based definition of college readiness, especially one that takes into account group differences in college completion. Broadly speaking, similar to the definition used by Conley (2007), this committee considers college readiness to be a student's preparedness to enroll in the degree-granting institution of their choice (2- or 4-year) without the need for remedial courses, to persist, and, ultimately, to earn a degree. Consistent with this definition, indicators of college readiness ideally should correspond to three stages of college: enrollment, persistence and first-year GPAs, and completion.

One set of college readiness metrics focuses on academic preparation, including students' test scores, GPAs, and advanced coursework (Adelman and Taylor, 1997; Glancy et al., 2014). Between-group differences along these dimensions are relatively straightforward to measure and have been well documented: they are discussed under Domain 2 (above) and Domain 6 (in Chapter 5). Among these indicators of academic readiness, unweighted high school GPA has the strongest evidence base as an indicator of readiness for college enrollment, college grades, persistence, and completion (Allensworth and Clark, 2018; Bowen, Chingos, and McPherson, 2009; Camara and Echternacht, 2000; Geiser and Santelices, 2007; Geiser and Studley, 2002; Roderick, Nagaoka, and Allensworth, 2006).

Focusing solely on academics, however, "masks the complexity of what it means to be ready to enroll and succeed in college and how [that readiness might differ] by student background and institutional characteristics" (Nagaoka, 2018, p. 2). Increasingly, research on what causes students to struggle or succeed in college has shown that readiness encompasses more than academic achievement (Braxton, 2000; Conley et al., 2014; Conley and French, 2014; Duckworth et al., 2007). Knowledge about college is another factor that influences college readiness, persistence, and completion. This knowledge includes an understanding of the logistics and processes (e.g., managing the application process, choosing the right college, and securing financial aid) and capabilities that facilitate social and academic success in college, such as having a growth mindset, self-regulation, social awareness, and a sense of belonging (Nagaoka, 2018). This suite of capabilities is especially important for students who come from families without college backgrounds.

In U.S. society, college knowledge as defined in these ways is unevenly distributed and varies by students' backgrounds and school environment (Conley, 2008). Without intervention, a lack of college knowledge "can discourage . . . and suppress the college aspirations of students, particularly first-generation college students, and students from racial/ethnic backgrounds who often find the college environment very different from their home communities" (Nagaoka, 2018, pp. 6-7).

One aspect of college knowledge, engagement in the college application process, is a strong predictor of college enrollment, particularly for low-income, minority students (Manski and Wise, 1983; Pallais and Turner, 2006; Plank, Deluca, and Estacion, 2008). It also shapes how students seek financial aid and pay for college. Applying for financial aid, particularly for low-income students, has been shown to predict enrollment in college (Lauff and Ingels, 2013; Roderick et al., 2008).

As a part of supporting the college choice and enrollment processes, it is possible to develop data systems that track students' progress on critical milestones that draw on college knowledge. These milestones include data on whether and where students submitted college applications and whether they successfully completed the Free Application for Federal Student Aid (FAFSA). Currently, some districts and states collect these types of information, and state-level FASFA completion data are available.[17] Schools should be cultivating and tracking other aspects of college knowledge, but those dimensions are not yet recommended for inclusion in a system of educational equity indicators because their measures are not well developed.

The committee also concludes it is important to track the paths that graduates pursue after they leave high school, including 2-year and 4-year programs, the military, employment, and unemployment. Even among high school graduates, there are large disparities in the subsequent paths chosen by students from various groups. These paths reflect different aspects of

---

[17]Information about FAFSA tracking can be found at https://www.collegeboard.org/membership/all-access/financial-aid/fafsa-tracking-pathway-college-affordability-and-student-access.

readiness and can lead to vastly different labor market and other long-term postsecondary outcomes. Gaps in pursuit of these outcomes can contribute to gaps in economic well-being of these groups as they become adults. The most fundamental construct to track is whether students are enrolled in any type of college at all (as opposed to entering the workforce or the military).

Figures 4-1 and 4-2 show the differences in postsecondary enrollment for students grouped by race and ethnicity and by SES, respectively.

Proposed Measures for Indicator 7

Postsecondary readiness can be measured by enrollment in higher education.

REVIEW OF EXISTING DATA SOURCES AND PUBLICATIONS

A key part of the committee's work was to investigate the potential usefulness of existing data systems and indicator reports for our proposed indicator set. Box 4-1 shows the criteria we used. Overall, although there is a wealth of information on pre-K to grade 12 education, the existing data and reports are not sufficient for the set of educational equity indicators as we have conceptualized them. Relevant information is scattered across multiple databases, which define some indicators and measure some constructs in different ways, do not provide any measures for some constructs, vary in data collection procedures, frequency, geographic detail, and coverage of student groups of interest, and are accessible through different agencies and organizations.

Tables 4-3, 4-4, and 4-5 summarize the potential data sources for each of the seven indicators and specific constructs for each indicator that we propose for Domains A, B, and C, respectively. The tables also summarize the extent to which data are ready to develop specific measures of each construct, and if not ready, what is needed. These tables draw on the information on existing data systems in Appendix A, on existing publications that include indicators of education equity in Appendix B, and on our assessment of data and methodological challenges and opportunities for educational equity indicators in Appendix C.

For these domains and indicators, the constructs and measures used generally pertain to students, categorized by groups of interest at the specific level of aggregation (e.g., school, school district, or state)—for example, the percentage of students who are chronically absent, separately for girls and boys. Note that immigration status of students is generally not available and that the measure of poverty status for students, if available, is usually in terms of whether the student is eligible for free or reduced-price school

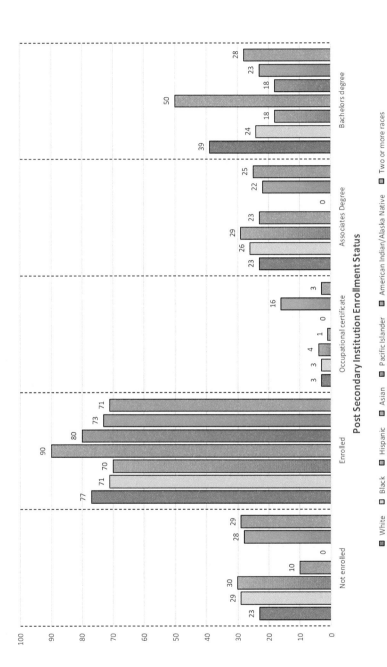

FIGURE 4-1 Percentage distribution of fall 2009 9th-grade students who had completed high school, by fall 2013 postsecondary enrollment status and by race and ethnicity: 2013.
SOURCE: Kena et al. (2016, p. 17).

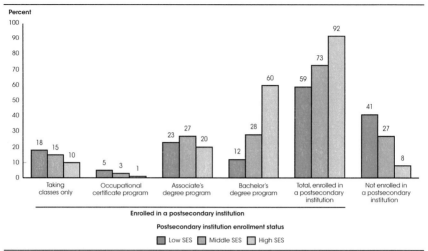

NOTE: SES was measured by a composite score based on parental education and occupations, family income, and school urbanicity in the student's 11th-grade year. The weighted SES distribution (weighted by W2STUDENT) was divided into five equal groups. Low SES corresponds to the lowest one-fifth of the population, and high SES corresponds to the highest one-fifth of the population. The three fifths in the middle were combined to form the middle SES category. Detail may not sum to totals because of rounding.
SOURCE: U.S. Department of Education, National Center for Education Statistics, High School Longitudinal Study of 2009 (HSLS:09), Base-Year, First Follow-up, 2013 Update, and High School Transcripts Restricted-Use Data File. See *Digest of Education Statistics 2015*, table 302.43.

FIGURE 4-2 Percentage distribution of fall 2009 9th-grade students who had completed high school, by fall 2013 postsecondary enrollment status and socioeconomic status (SES): 2013.
SOURCE: Kena et al. (2016, p. 18).

lunch. As discussed in Appendix C, this measure is less and less appropriate as a measure of poverty status, so work will be needed to develop an appropriate measure from data that are feasible to collect.

BOX 4-1
Criteria for Committee Review of
Existing Data Systems and Publications

1. Published on a regular, frequent basis—at least annually.
2. Available for subnational geographic areas, including states, school districts, and, ideally, schools or school attendance areas, as appropriate.
3. High-quality when assessed on measures of nonsampling error (e.g., accurate reporting of student enrollment) and on measures of sampling error (for survey-based data).
4. Available for groups of students of interest for educational equity (see Chapter 2), as defined by race and ethnicity, gender, family income (or equivalent measure of socioeconomic resources), disability status, immigrant status, and English-language capability.
   a. For immigrant students, indicative of time of entry into the United States to appropriately include/exclude them in equity indicators (e.g., exclude from a high school graduation measure if they arrived only a year before graduation).
   b. For English-language learners, when possible, indicative of the number of years spent in an English-learner program, whether a student waived out of English-learner instruction, and time and type of reclassification to English-proficient status.
5. Measures contextual factors, such as neighborhood income and family type composition for student groups of interest (see Chapter 3).[a]
6. Measures equity in students' educational outcomes for student groups of interest (see text above and tables below).
7. Measures equity in school-provided opportunities to learn for student groups of interest (see Chapter 5).
8. Constructed in a manner that is intelligible to users of varying levels of analytic sophistication.
9. Constructed so that it is difficult to "game" the indicator to make a school district or school appear to be more equitable than it is.
10. Feasible to produce on a timely basis (i.e., soon after the underlying data are available).

_____

[a] Although we do not propose indicators of context, they would be critical to inform efforts of school systems to work with other sectors to combat root causes of poverty and other factors that adversely affect students' educational attainment.

TABLE 4-3 Potential Data Sources and Measures for Domain A, Kindergarten Readiness

Constructs	Source (Characteristics)
Indicator 1: Disparities in Academic Readiness	
Reading/Literacy Skills Numeracy/Math Skills	**Source:** NCES ECLS-K:2011 **Frequency:** One time **Geographic detail:** Nation (sample is too small for finer detail) **Student group detail:** Race/ethnicity (based on parents), gender, whether English spoken at home, whether family received public assistance **Possible measures:** Average scale score or percentage of students within specified range of average on reading and math assessments conducted at beginning of kindergarten school year **Future potential:** Use tested assessments in ECLS-K:2011 to develop assessments that are age appropriate and feasible for schools to administer at scale, nationwide and annually
Indicator 2: Disparities in Self-Regulation and Attention Skills	
Self-Regulation Skills Attention Skills	**Source:** NCES ECLS-K:2011 **Frequency:** One time **Geographic detail:** Nation (see Indicator 1, above) **Student group detail:** Same as Indicator 1, above **Possible measures:** Average scale scores on direct assessments of social skills and learning behaviors, or on teachers' assessments of same, conducted after first month of kindergarten **Future potential:** See Indicator 1, above, but the road to feasible, streamlined assessments for nationwide, annual use is likely even more challenging, given the need for observational data

NOTES: ECLS-K, Early Childhood Longitudinal Survey-Kindergarten; NCES, National Center for Education Statistics.

TABLE 4-4 Potential Data Sources and Measures for Domain B, K–12 Learning and Engagement

Constructs	Source (Characteristics)
Indicator 3: Disparities in Engagement in Schooling	
Attendance/ Absenteeism (a proxy construct for engagement or lack of engagement)	**Source:** ED*Facts* (as part of ESSA reporting requirements)
	Frequency: Annual
	Geographic detail: Nation, states, districts, schools
	Student group detail: Race/ethnicity, gender, English-language status, disability status
	Grade/level detail: Elementary, middle, secondary, other; schools can be classified by percent students eligible for free or reduced-price lunch
	Possible measures: Percent students chronically absent above a specified threshold (10 percent or more school days in ED*Facts*)
Indicator 4: Disparities in Performance on Coursework	
Success in Classes (record of passing courses)	**Source:** Transcript studies conducted in NAEP and NCES longitudinal surveys
	Frequency: Periodic
Accumulating Credits (being on track to graduate)	**Geographic detail:** Nation (samples too small for finer detail)
	Student group detail: Race/ethnicity, gender, English-language status, disability status, eligible/not eligible for free or reduced-price lunch
Grades, GPAs	**Grade/level detail:** Transcripts collected at end of high school
	Possible measures: Percent high school seniors above a specified threshold passing all courses; percent seniors above a specified threshold having enough credits to graduate; average GPA for seniors
	Future potential: Construct from SLDS as more states develop them in a comparable manner and provide access for statistical purposes
	Comment: The CRDC has students passing algebra I in grades 8 and 9-10 for nation, states, districts, and schools for student groups defined by race/ethnicity, gender, English-language status, and disability status, collected biannually

continued

TABLE 4-4 Continued

Constructs	Source (Characteristics)
Indicator 5: Disparities in Performance on Tests	

Constructs	Source (Characteristics)
Achievement in Reading, Math, and Science Learning Growth in Reading, Math, and Science Achievement	**Source (1):** Main NAEP **Frequency:** Biannual for reading and math (every 4 years for 12th graders); periodically for science **Geographic detail:** Nation, states, some large city districts **Student group detail:** Race/ethnicity, gender, eligible/not eligible for free and reduced-price lunch **Grade/level detail:** 4th, 8th, 12th grades (4th, 8th grades for science) **Possible measures:** Percent students achieving at or above proficient level, or average scale scores; average of per student change in scale scores over time **Source (2):** EDFacts (as part of ESSA reporting requirements) **Frequency:** Annual **Geographic detail:** Nation, states, districts, schools **Student group detail:** Race/ethnicity, gender, English-language status, disability status, economically disadvantaged (typically eligible/not eligible for free or reduced- price lunch) **Grade/level detail:** Each grade from 3 to 8 and once in high school (reading and math); one grade in grades 3-5, 6-9, and 10-12 (science) **Possible measures:** Percent students achieving at or above a specified achievement level (e.g., the middle value of the levels, which may be from 3 to 6 in various states); percent increase in students achieving at or above a specified level **Comment:** States do not use the same assessments or the same number or definitions of achievement levels, so a method is needed to make results comparable—SEDA has developed correction factors based on NAEP (see Appendix C)

NOTES: CRDC, Civil Rights Data Collection; ESSA, Every Student Succeeds Act of 2015; GPA, grade point average; NAEP, National Assessment of Educational Progress; NCES, National Center for Education Statistics; SEDA, Stanford Education Data Archive; SLDS, Statewide Longitudinal Data System.

TABLE 4-5 Potential Data Sources and Measures for Domain C, Educational Attainment

Constructs	Source (Characteristics)
Indicator 6: Disparities in On-Time Graduation	
On-Time Graduation	**Source:** ED*Facts* (as part of ESSA reporting requirements)
	Frequency: Annual
	Geographic detail: Nation, states, districts, high schools
	High school student group detail: Race/ethnicity, gender, English-language status, disability status, economically disadvantaged (typically eligible/not eligible for free or reduced-price lunch)
	Possible measure: Adjusted Cohort Graduation Rate (ACGR)
	Comment: The ACGR is widely accepted for measuring on-time high-school graduation (see Appendix C)
Indicator 7: Disparities in Postsecondary Readiness	
Enrollment in College, Entry into the Workforce, or Enlistment in the military (after completion of high school)	**Source:** American Community Survey (ACS)
	Frequency: Annual
	Geographic detail: Nation, states, districts (5-year averages)
	High school graduate group detail: Race/ethnicity, gender, English-language status (based on language spoken in the home), disability status (limited number of conditions), poverty status
	Possible measures: Percent high school graduates ages, say, 18-21, in college, the workforce, or the military; percent all young adults ages, say, 18-21, in college, the workforce, or the military (this measure includes high school dropouts in the denominator)
	Future potential: Construct from SLDS as more states develop them in a comparable manner, follow graduates beyond high school, and provide access for statistical purposes
	Comment: ACS data cannot readily be linked to the graduate or young adult's school district (some linkage could be possible with a question on whether one lived in the same house a year ago); the SLDS would obviate this problem to the extent that graduates can be followed up

NOTES: ESSA, Every Student Succeeds Act of 2015; SLDS, Statewide Longitudinal Data System.

5

Indicators of Disparities in
Access to Educational Opportunities

As described in Chapter 4, we propose indicators that fall into two categories: indicators of disparities in students' educational outcomes and indicators of disparities in students' access to educational resources and opportunities. This chapter addresses the second category of indicators—those related to opportunities and resources. Table 5-1 shows the indicators we propose.

The five sections in this chapter cover each of the four domains related to opportunities, usually termed inputs or resources, and a fifth on the availability of data and measures for those four domains. In Appendix C the committee provides illustrations of the data sources and methods that could be used to develop appropriate measures for our proposed indicators.

DOMAIN D: EXTENT OF RACIAL, ETHNIC,
AND ECONOMIC SEGREGATION

School segregation—both racial and economic—poses one of the most formidable barriers to educational equity. Under conditions of racial and economic segregation, black, Hispanic, and low-income students disproportionately attend schools with high concentrations of other black, Hispanic, and low-income students, while students from white and nonpoor families attend schools with other white children and children from families with more resources.

Segregation limits opportunities for children of all backgrounds to develop and enhance important life skills, such as the ability to interact effectively with diverse groups (Pettigrew and Tropp, 2006). Attending

TABLE 5-1 Proposed Indicators of Disparities in Access to Educational Opportunities

DOMAIN	INDICATORS	CONSTRUCTS TO MEASURE
D Extent of Racial, Ethnic, and Economic Segregation	8 Disparities in Students' Exposure to Racial, Ethnic, and Economic Segregation	Concentration of poverty in schools Racial segregation within and across schools
E Equitable Access to High-Quality Early Learning Programs	9 Disparities in Access to and Participation in High-Quality Pre-K Programs	Availability of licensed pre-K programs Participation in licensed pre-K programs
F Equitable Access to High-Quality Curricula and Instruction	10 Disparities in Access to Effective Teaching	Teachers' years of experience Teachers' credentials, certification Racial and ethnic diversity of the teaching force
	11 Disparities in Access to and Enrollment in Rigorous Coursework	Availability and enrollment in advanced, rigorous course work Availability and enrollment in Advanced Placement, International Baccalaureate, and dual enrollment programs Availability and enrollment in gifted and talented programs
	12 Disparities in Curricular Breadth	Availability and enrollment in coursework in the arts, social sciences, sciences, and technology
	13 Disparities in Access to High-Quality Academic Supports	Access to and participation in formalized systems of tutoring or other types of academic supports, including special education services and services for English learners

TABLE 5-1 Continued

DOMAIN	INDICATORS	CONSTRUCTS TO MEASURE
G Equitable Access to Supportive School and Classroom Environments	14 Disparities in School Climate	Perceptions of safety, academic support, academically focused culture, and teacher-student trust
	15 Disparities in Nonexclusionary Discipline Practices	Out-of-school suspensions and expulsions
	16 Disparities in Nonacademic Supports for Student Success	Supports for emotional, behavioral, mental, and physical health

integrated and racially and culturally diverse schools can help to increase students' comfort with diversity and understanding of different perspectives, which has been associated with improvements in critical thinking, communication, and problem solving (Kurlaender and Yun, 2005, 2007). Integrated schools have been shown to be better for all students. There is evidence that racially integrated schools are associated with greater life outcomes for all students, including higher college enrollment and success, higher lifetime earnings, a more diverse circle of friends and living arrangements in adulthood, and the important career skill of working with people from diverse backgrounds (Philips, Gormley, and Lowenstein, 2009; Siegel-Hawley, 2012; Wells, Fox, and Cordova-Cobo, 2016).[1]

Indicator 8: Disparities in Students' Exposure to Racial, Ethnic, and Economic Segregation

Racial segregation—as measured by how evenly black and Hispanic students are distributed among U.S. public schools and public school districts—continues to be a problem, and recent data show that it has increased in recent decades. Measured by the proportion of schools clas-

---

[1]For additional information, see https://www.civilrightsproject.ucla.edu/research/k-12-education/integration-and-diversity/brown-at-62-school-segregation-by-race-poverty-and-state/Brown-at-62-final-corrected-2.pdf; http://www.centerforpubliceducation.org/system/files/School%20Segregation%20Full%20Report_0.pdf; https://www.epi.org/publication/the-racial-achievement-gap-segregated-schools-and-segregated-neighborhoods-a-constitutional-insult/.

sified as "high-minority" schools (75% or higher black and Hispanic), racial segregation increased from 9 percent in 2000–2001 to 16 percent in 2013–2014 (U.S. Government Accountability Office, 2016). Some of the increase is due simply to the increasing proportion of black and Hispanic students in U.S. schools. That is, when measured by statistics describing how evenly students of different races are distributed among schools, given the current national racial composition, average school segregation levels have been relatively unchanged over the past few decades (see Reardon and Owens, 2014). The lack of progress in reducing these levels of racial segregation is due to several factors, especially the barriers to desegregation created by school district boundaries and by post-1973 changes in Supreme Court rulings. Those rulings limited the circumstances in which courts can impose desegregation orders, as well as the circumstances in which districts may voluntarily adopt race-sensitive policies to reduce segregation (Black, 2017; Yudof et al., 2011; Orfield et al., 2016).

Economic segregation—as measured by how evenly poor and nonpoor students are distributed among U.S. public schools and public-school districts—has also risen steadily since the 1970s partly as a result of increasing income inequality and a rise in income-based housing segre-gation, especially among families with school-aged children. Sometimes referred to as "double segregation" (Orfield et al., 2016) or "hyper-segregation" (George and Darling-Hammond, 2019), these factors have led to circumstances in which black and Hispanic students are more likely than their white and Asians peers to be in schools with high levels of students from economically disadvantaged families. These patterns have been shaped by federal, state, and local schooling *and* housing policies, by racial and economic inequality, and by a history of housing discrimination (Orfield, 2013; Rothstein 2015). Black and Hispanic students are dispro-portionately likely to be low-income themselves, but they are even more disproportionately likely to be enrolled in schools with large proportions of low-income students (see Figure 5-1).

Research shows that schools that serve a majority of students from economically disadvantaged communities often lack the human, material, and curricular resources to meet their students' academic and socioemo-tional needs. Consequently, their students have unequal access to the full suite of learning opportunities and resources that can promote their success and are available to children from wealthier families (Owens, Reardon, and Jencks, 2016). Indeed, school poverty rates are associated with key measures of school quality that affect learning and achievement (Bohrnstedt et al., 2015; Clotfelter et al., 2007; Hanushek and Rivkin, 2006). In one example elaborated elsewhere in this chapter, students in high schools serving high populations of students of color (defined as schools where at least 75 percent of students are black or Hispanic) or

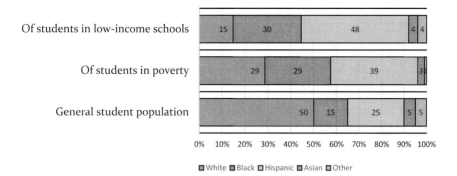

FIGURE 5-1 Comparison of proportion of students in poverty and proportion of students in low-income schools, by race, 2014.
SOURCE: Wagner (2017, p. 14). Reprinted with permission from *School Segregation Then & Now: How to Move toward a More Perfect Union*, Copyright 2017, National School Boards Association. All rights reserved.

high populations of students in poverty[2] are less likely to have access to courses needed to prepare them for college and careers. Schools serving high concentrations of students in poverty offer fewer advanced placement (AP) courses and gifted and talented education programs (U.S. Government Accountability Office, 2016).

There are many dimensions of segregation that might be included in an indicator of educational equity. Potential indicators might include measures of racial or economic segregation and measures of residential or school segregation. There are also several ways to measure segregation. Exposure is the extent to which students of a given race or ethnicity attend schools with students of another race or ethnicity, and isolation is the inverse of exposure. Unevenness is the extent that students are evenly distributed by race or ethnicity among schools within a district or other region. If all schools have the same racial and ethnic composition as one another (and therefore the same as that of the district or region), then there is evenness (Massey and Denton, 1988).

---

[2]The definition is based on identifying low-income students as those eligible for free or reduced-price lunch under the National School Lunch Program, sorting K–12 schools according to the percentages of eligible students, separating out high schools, and dividing the high schools into quintiles, which we reference throughout the report (ExcelinEd, 2018).

Recent research has shown that the dimension of segregation most strongly associated with achievement is the racial difference in exposure to poor schoolmates (Reardon, 2016). That is, in places where black or Hispanic students attend schools with higher poverty rates than do white students, the white-black and white-Hispanic test score gaps are larger, on average. This measure—which captures the combination of racial and economic segregation—was more predictive of achievement gaps than measures of residential segregation, of racial segregation alone, or of economic segregation alone. These findings suggest that both economic segregation and racial segregation are harmful to academic achievement, and importantly, racial segregation is harmful because it leads to the economic segregation of white and nonwhite students.

Proposed Measures for Indicator 8

The committee proposes an indicator that is focused on the difference in poverty rates in schools attended by poor and nonpoor students, by students from different racial groups, by English-language learners, and by students with immigrant or foreign-born parents. Such an indicator is readily interpretable and has historically been relatively straightforward to measure so long as school districts report reliable school-level counts of students eligible for free or reduced-price meals.[3] The committee also proposes an indicator of racial segregation within and across schools.

DOMAIN E: EQUITABLE ACCESS TO HIGH-QUALITY EARLY LEARNING PROGRAMS

Early childhood learning is a strong predictor of kindergarten readiness and "one of the most common and policy relevant out-of-home experiences" that young children can have (Magnuson, 2018, p. 8). However, there are sizable differences in the availability of licensed early learning programs and in enrollment: those differences are between children growing up in disadvantaged circumstances and their more advantaged peers. And that availability gap is compounded by a corresponding disparity in the quality of programs that are available to children from families with different income levels.[4]

---

[3]We note, however, that some 20 percent of schools now provide free meals to all students, regardless of their family income, under the community eligibility provision of the Healthy, Hunger-Free Kids Act of 2010. As a result of this provision, in many cases these schools no longer report accurate counts of students who are poor, complicating the measurement of differences in exposure to schoolmates who are poor. In order to provide accurate measures of segregation for this measure, it would be necessary to have accurate measures of school poverty that are comparable across schools, districts, and states (see Appendix C).

[4]See https://nces.ed.gov/programs/digest/d17/tables/dt17_202.10.asp.

Indicator 9: Disparities in Access to and Participation in High-Quality Pre-K Programs

Participation in an early learning program is essential for ensuring that children develop the behaviors and competencies they will need to do well in kindergarten. Research suggests that participating in preschool programs for at least 2 years prior to entry to kindergarten is beneficial for all children (see Magnuson, 2018). Investments by federal, state, and local programs have increased considerably in the past 30 years in efforts to reduce enrollment gaps and improve access to high-quality early childhood education for disadvantaged populations. Nevertheless, children from lower income families, families with lower levels of educational attainment, and households in which the parents are not proficient in English—children who could benefit most from programs—are often the least likely to enroll in them (Table 5-2 shows these data). In particular, we note that:[5]

- 56.4 percent of Latino 3- to 5-year-olds were not enrolled, compared with 43.2 percent of white children.
- 50.5 percent of children with foreign-born parents were not enrolled, compared with 45.3 percent of children with native-born parents.
- 60 percent of children whose parents did not graduate high school were not enrolled, compared with 56.1 percent of children whose parents are high school graduates and 36.4 percent of children whose parents have a college degree.

While participation in any preschool program can be beneficial to children, participation in high-quality programs is even more important. At present, while measures of quality exist, they vary in terms of what is measured and how it is measured. Dimensions of quality include classroom resources; curriculum; interaction quality between teachers and children; and teachers' credentials and experience.

Programs also vary in their complexity to implement, their feasibility, and their costs, as well as the validity of the rating systems. For instance, since all states have regulations and licensing standards for child care providers, one might think that simply being licensed could be the signal of program quality. However, state licensing regulations and standards vary widely, and not all of them address issues of quality as they relate to children's development and learning. For example, many are focused on protecting children from harm, such as by mitigating risks from inadequate

---

[5]Child Trends (2015) has tabulated these percentages since 1994, and the pattern is similar across the years.

TABLE 5-2 Enrollment of 3- to 5-Year-Olds in Preschool and Pre-K Programs by Race and Ethnicity, Parental Education, Immigrant Status, and Household Income, 2013: in Percent

Population Group	Extent of Participation in Preschool, Pre-K Program		
	Full Day	Part Day	None
Race and Ethnicity			
White, Non-Hispanic	25.2	31.6	43.2
Black, Non-Hispanic	38.6	17.6	43.7
Hispanic	22.0	21.6	56.4
Parental Education			
Less than a High School Degree	20.3	19.7	60.0
High School Degree or Equivalent	24.6	19.4	56.1
Some College or Technical or Vocational Degree	23.4	26.6	50.0
Bachelor's Degree or Higher	30.1	33.6	36.4
Immigrant Status			
Both Parents Native Born	27.1	27.5	45.3
One or Both Parents Foreign Born	23.9	25.6	50.5
Household Income			
Less than $15,000	23.8	22.7	53.5
$15,000–$29,999	21.9	22.9	55.2
$30,000–$49,999	24.2	23.2	52.6
$50,000–$74,999	24.0	27.9	48.1
$75,000 and over	21.7	32.4	35.9

NOTE: The data exclude children aged 3-5 who are enrolled in kindergarten or elementary school.
SOURCE: Information from Child Trends (2015).

supervision, poor building and hygiene standards, and unsafe practices (Workman and Ullrich, 2017). They may also specify education requirements for preschool teachers. Many stipulate fundamental components necessary for operation, but do not address the comprehensive developmental and learning needs of preschoolers.

Observational measures of classroom instructional quality are somewhat stronger predictors of children's learning, although observations are expensive and labor intensive, and existing research shows that the associations are small (Burchinal, Kainz, and Cai, 2011).

Although challenging to implement, quality rating and improvement systems are available and have been implemented in all states except Mississippi (see Box 5-1). More than half include an observational rating system designed to measure caregiver responsiveness and program stimulation (Tout et al., 2010).

The National Institute for Early Education Research publishes state report cards that rate early childhood programs on the basis of 10 criteria.[6] The criteria are consistent across states and thus allow comparisons, but the information about program quality is limited.

These rating systems can be used at the state level to characterize the quality of a state's early childhood education programs and possibly also reported at the school district level (or for other regions in a state). Because these rating systems differ from state to state, however, they cannot be used in a national indicator system without a major caveat: the data will only indicate provider quality in relation to what each state defines as adequate, not in relation to a national quality standard—because none exists in federal policy or in a consensus among researchers. However, even if a common measure of quality existed, it would not be useful for measuring disparities unless data on the demographics of the children enrolled at each center was known. Coordinated decision making would be necessary to select and refine a standard measure of program quality in a national indicator system, together with appropriate data collection systems.

Proposed Measures for Indicator 9

The committee proposes that Indicator 9 be measured in two ways: (1) the availability of licensed early childhood education programs and (2) enrollment in these programs. We would have liked to include access to high-quality programs as an indicator, but we cannot do so because of the inconsistencies and variability in measuring quality across states and other jurisdictions.

DOMAIN F: EQUITABLE ACCESS TO HIGH-QUALITY CURRICULA AND INSTRUCTION

Interaction between students and their teachers—through curriculum, coursework, and instruction—is at the heart of education. Students' exposure to a rich and broad curriculum, challenging coursework, and inspired teaching is therefore vital for their learning and development. There is no

---

[6]See http://nieer.org/state-preschool-yearbooks.

BOX 5-1
How States Currently Measure Quality in Early Childhood
Education Programs

States use one or more of four measures for early childhood education programs: environment rating scales, the Classroom Assessment Scoring System (CLASS), national accreditation, and a quality rating and improvement system.

Environment Rating Scales: The Early Childhood Environment Rating Scale (ECERS) for children ages 3-5, the Infant/Toddler Environment Rating Scale, and the Family Child Care Environment Rating Scale, are standardized tools used to measure process quality at the classroom level. The measures contain multiple items on which programs are rated, organized into seven subscales. These subscales include ratings of the space and furnishing, personal care routines, the activities and interactions that take place in the classroom, and how the program engages with families. Ultimately, these tools are designed to assess the various interactions that occur in the learning environment—between staff and children, among children themselves, and among children and materials and activities—and the structures that support these interactions, such as the space and the schedule.

CLASS: CLASS is an observation tool that assesses the interactions between teachers and children that affect learning and development. CLASS has separate scales for different age groups, reflecting the differences in how infants, toddlers, and preschoolers learn. The infant observation has just one domain while the pre-K observation has three domains. The observation assesses the quality of relationships, routines, the organization of the physical environment, and the way language is used and interactions are facilitated to prompt children to think critically.

---

widespread agreement on which specific elements of curriculum, coursework, and teaching matter for student outcomes. Most of the research base is inadequate to support causal inferences about the relationships between these factors and student outcomes.

But there is evidence that these core elements are not distributed in an equitable way—in relation to either proportionality or need. Excellence in academic programming and resources needs to include not only equitable access to AP courses and other advanced coursework, but also meeting the academic needs of students on the other end of the achievement distribution. The adequacy of formal academic supports for students who are struggling to achieve is at least as important as fair access to enrichment opportunities for students at the top.

In the absence of research-based clear causal links of specific curricula, instructional practices, and courses with desired student outcomes, we are

National Accreditation: Accreditation is a voluntary process that programs can use to help improve their level of quality and to demonstrate to families—both of children currently enrolled and prospective enrollees—that the program has gone above and beyond what is required by state regulation and achieved a specified level of quality. To achieve these accreditations, programs need to engage in extensive self-study and go through an external validation process. While these accreditations do differ, most contain a number of common standards. For example, they generally include standards related to the learning environment, teacher and child interactions, staff qualifications, professional development, and family engagement, among others.

Quality Rating and Improvement Systems: All states either have a quality rating and improvement system, a pilot for one, or are in the planning process for one. These systems are designed to assess, improve, and communicate the level of quality in early childhood education settings. Programs are assessed on multiple elements and receive a rating reflecting their level of quality, usually on a scale of 1 to 3 or 1 to 5. There is no one system in use across the United States; every state has a unique system reflecting its own priorities and context. Many quality rating and improvement systems do include environmental observations, such as ECERS or CLASS, as part of their assessment activities, and these scores factor into the overall rating. Other elements of the rating might include family engagement activities, child assessments, and program management. Many systems also waive some requirements for programs with national accreditation or automatically grant programs a certain rating as a result of their national accreditation.

SOURCE: Information from Workman and Ullrich (2017).

proposing proxies based on the available research and the committee's collective judgment.

Indicator 10: Disparities in Access to Effective Teaching

This committee is not the first to recognize the important role teachers play in promoting student learning. Indeed, there is widespread agreement that teachers are the most important in-school factor contributing to student outcomes (Aaronson, Barrow, and Sander, 2007; McCaffrey et al., 2004, 2009; Nye, Konstantopoulos, and Hedges, 2004; Rivkin, Hanushek, and Kain, 2005; Rockoff, 2004; Sanders and Rivers, 1996). These studies, which use value-added modeling (VAM) to estimate the effect of a teacher on students' achievement test scores, have consistently found that the effects of individual teachers on their students' achievement is substantial and persistent. Although some scholars have identified limitations in these models (e.g., Rothstein, 2017), the large body of work on VAM has led to widespread support for claims about the importance of teachers. This is also a point of consensus among policy influencers (U.S. Department of Education, 2013).

While teacher effects on student learning have been extensively studied, recent work also finds that teachers have meaningful effects on engagement, behavioral skills, and other outcomes (e.g., Blazar and Kraft, 2017; Jackson, 2014; Ladd and Sorenson, 2017). This work also finds that teacher effects are multidimensional—those teachers who produce the best gains in student achievement do not necessarily produce the best gains in other outcomes (Blazar and Kraft, 2017; Jackson, 2014; Jennings and DiPrete, 2010; Kraft, 2019). Despite this knowledge, the field lacks useful measures of what makes teachers effective, especially with underserved student populations. For the measures that do exist, they are not available at scale for use in an indicator system.

The committee has chosen not to include other, commonly used measures of teaching effectiveness that focus specifically on teachers' pedagogical skills or instructional practices. These strategies include observational measures, in which the supervisor or other expert rates teachers' pedagogical quality according to key dimensions thought to characterize effective instruction, and student survey measures, in which students rate teachers' instructional quality along similar key dimensions. There is a wealth of information about these measures, including guidance for developing, implementing, and using these kinds of measures (see, e.g., Cantrell and Kane, 2013; Gitomer and Zisk, 2015; Kane, Kerr, and Pianta, 2014). We wholeheartedly endorse use of these measures for local efforts to improve teaching, but we do not propose their use as indicators because they differ so much across jurisdictions and because they generally have not been validated for use in a large-scale indicator system.

We also have not included measures of teaching effectiveness as estimated through statistical modeling. These estimates can take a variety of forms, of which VAM is the most common. They are widely used as part of some states' teacher evaluation systems, but this use is controversial. As the disagreements among researchers and stakeholders in the teacher evaluation field remain unresolved, the committee did not attempt to come to consensus about any of these issues.[7] Any of these three measures could be used as an indicator in an educational equity system, but they could only provide information at the state or local level. We instead offer three other

---

[7]Describing the disagreements about value-added methods is beyond the scope of this report. For reviews of this issue, see, for example, American Educational Research Association (2015); American Statistical Association (2014); and National Research Council (2010).

measures: teachers' years of experience, teachers' qualifications for the subjects they teach, and teacher diversity.

Teachers' Years of Experience

Group differences in exposure to novice teachers are important to consider from an equity perspective. Overall, 5 percent of the nation's 3 million teachers (full-time equivalent) are in their first year of teaching. However, schools serving the highest percentages of black and Latino students in their school district are more likely to employ teachers who are newest to the profession. These schools reported 6 percent of their teaching staff as being in their first year of teaching in any school, compared with 4 percent in schools with the lowest percentage (bottom 20%) of black and Latino students in their districts (Rahman et al., 2017). Of the nearly 5 million English learners nationwide, 3 percent attend schools where more than 20 percent of teachers are in their first year of teaching, compared with 2 percent of non-English learner students.

Teachers' Qualifications for the Subjects They Teach

Teacher certification—itself a proxy for teachers having the relevant knowledge and skills to teach effectively—is not strongly associated with desired outcomes (Hanushek and Rivkin, 2010; Aaronson, Barrow, and Sander, 2007; Kane, Rockoff, and Staiger, 2008; Rockoff, 2004). However, direct measures of teachers' knowledge seem predictive of student performance in some areas, especially math and science (e.g., Baumert et al., 2010; Hill, Rowan, and Ball, 2005; Sadler et al., 2013).

In these studies, teachers' knowledge is typically measured by researcher-developed tests of teachers' content knowledge or pedagogical content knowledge (e.g., Baumert et al., 2010; Hill, Rowan, and Ball, 2005; Sadler et al., 2013), though it is also possible to use existing assessments, such as college entrance exams or teacher certification exams, for this purpose. To date, however, the research on teacher knowledge has been limited to a small number of studies that do not address every grade and subject, and there is insufficient evidence regarding the appropriateness of existing knowledge tests for use in an indicator system.

In the absence of more refined measures, we propose including teacher certification in the system of equity indicators. Despite the lack of evidence that certification affects student achievement, it does provide a signal that teachers have attained a basic level of knowledge and skills, so students whose teachers lack certification might be at a disadvantage. Researchers have identified systematic between-group differences in access to certified teachers.

- Nationwide, 97 percent of teachers met all state certification or licensure requirements in the 2011-2012 school year. However, the Civil Rights Data Collection (CRDC) reveals that nearly 0.5 million students are enrolled in schools in which 60 percent or fewer of the teachers met all state certification requirements (U.S. Department of Education Office of Civil Rights, 2014).
- Racial disparities exist in students' access to certified teachers: black students are more than four times as likely, and Latino students twice as likely, as their white peers to attend schools where 20 percent or more of their teachers have not yet met all state certification and licensing requirements. Nearly 7 percent of black students attend schools in which more than 20 percent of the teachers have not yet met all state certification and licensing requirements, compared with 3.7 percent of Latinos and slightly less than 2 percent of white students.

Teacher Diversity

The racial and ethnic distribution of students enrolled in public school has been gradually changing over the past few decades. As of 2015, a bare majority of public school students across the country were nonwhite; 49 percent were white.[8] Specifically, public school students were 16 percent black, 26 percent Hispanic, 5 percent Asian, and 5 percent other or two or more races. By 2027, the population of students enrolled in public schools is projected to become even more diverse: 45 percent white, 15 percent black, 29 percent Hispanic, 6 percent Asian, 1 percent Native American, and 4 percent of two or more races.[9]

In contrast, there is far less diversity among teachers: in 2015, just 20 percent of teachers were nonwhite: 7 percent were black, 9 percent were Hispanic, 2 percent were Asian, and 2 percent were other or two or more races. In addition, nonwhite teachers are highly concentrated in certain areas: in 2011 an estimated 40 percent of schools had no nonwhite teachers, meaning nonwhite students in those schools might never experience a teacher of their own race or ethnicity (Bireda and Chait, 2011).

There is growing and compelling evidence that teacher-student racial match has important effects on student outcomes. These match effects appear on both short-term outcomes, such as student test scores and academic attitudes (Dee, 2004; Egalite and Kisida, 2018; Egalite, Kisida, and Winters, 2015; Goldhaber and Hansen, 2010), and long-term outcomes, such as dropping out of high school (Gershenson, Jacknowitz, and

---

[8]See https://nces.ed.gov/programs/digest/d17/tables/dt17_203.50.asp?current=yes.
[9]See https://nces.ed.gov/programs/digest/d17/tables/dt17_203.50.asp?current=yes.

Brannegan, 2017). They also are found on nonacademic outcomes, such as student disciplinary outcomes (Holt and Gershenson, 2017; Lindsay and Hart, 2017).

These effects are not small. For instance, in the Tennessee Student/ Teacher Achievement Ratio (STAR) class-size study, the effects of student-teacher racial/ethnic match were as large as the effects of small classes themselves (Dee, 2004).[10] In other words, students who were randomly assigned to a small class received an increase in achievement, and students (both black and white) who were randomly assigned to an own-race teacher received an equally large increase in achievement. Depending on model specification, those increases were not small in magnitude, ranging from 5 to 8 percentile points on a nationally normed achievement test. In terms of longer-term outcomes, black students randomly assigned a black teacher in the STAR study were 7 percent more likely to graduate from high school and 13 percent more likely to aspire to college than black students who were not randomly assigned a black teacher (Gershenson et al., 2018).

Given the persistent racial achievement gaps and demographic shifts in the United States, there is a new urgency to understand this phenomenon. Though more research is needed, the existing evidence suggests that the diversity of a school's teaching staff and its match to the student body merits inclusion in a system of equity indicators.

Proposed Measures for Indicator 10

Measures for Indicator 10 should include teachers' years of experience; teachers' credentials for the subjects they teach; and diversity of the teaching force to which students are exposed.

Indicator 11: Disparities in Access to and Enrollment in Rigorous Coursework

Research has long shown that differences in exposure to challenging courses and instruction contribute to disparities in educational outcomes by race, ethnicity, and socioeconomic status (Gamoran, 1987; Gamoran and Mare, 1989; Oakes, 1985). These disparities can result from factors that differentially affect entire schools (between-school differences) or that differentially affect specific groups of students within a school (within-school differences). The former can arise from circumstances such as residential segregation (see Domain D, above) where access to opportunities and resources differs for schools with high and low concentrations of students in poverty. The latter can arise when students are assigned to coursework

---

[10]For details, also see https://dataverse.harvard.edu.

using methods commonly known as "tracking," such as assigning students to the college preparatory track or the general education track or using within-class ability groupings. While tracking may be well-intentioned as an instructional strategy, it can also mean that student groups are disproportionately placed in courses of differing levels of rigor, even if they have similar levels of ability or similar prior academic performance (Mickelson, 2005; Orfield and Lee, 2005).

In addition to organizational policies, various factors influence the within-school distribution of opportunities to students from different groups. Examples include teacher subjectivity (Dougherty et al., 2015; Grissom and Redding, 2016; Thompson, 2017); parents' efforts on students' behalf (Lewis and Diamond, 2015); and having a critical mass of prepared students (Iatarola, Conger, and Long, 2011). Counteracting some of these factors may require quite specific, tailored, school-level actions.

Inequitable access to rigorous coursework may be an especially serious issue for students with disabilities and English learners. Though federal law encourages the inclusion of students with disabilities in general education classrooms, they are often excluded from advanced or honors coursework.[11] Even in general education classrooms, students with disabilities experience less time learning content in the grade-level standards, less instructional time, and less content coverage than their nondisabled peers (Kurz et al., 2014). Perhaps as a result, students with disabilities are much less likely than their nondisabled peers to expect to enroll in postsecondary education or take the necessary entrance exams (Lipscomb et al., 2017).

For English-language learners, the findings are similar: they have less opportunity to learn rigorous content in the classroom, often due to language barriers between them and their teachers (Abedi and Herman, 2010). They are also less likely than their English-language fluent peers to be exposed to the regular academic curricula in high school (Callahan and Shifrer, 2016; Umansky, 2016). Again perhaps because of these disparities, English learners are far less likely than native English speakers to subsequently attain college degrees (Kanno and Cromley, 2013).

Whether these disparities are caused by within-school or between-school factors, they can contribute to disparities in other desired outcomes. For example, differential access to prerequisite courses in middle school and in the early years of high school leaves many students ineligible or unprepared to take advanced courses. Some state university systems have course-taking requirements for entry that are more difficult to meet in schools that do not routinely offer all the necessary courses (Gao, 2016). In California, for example, inequities in course access and completion have resulted in large gaps in readiness for entrance to a college in either the

---

[11]See https://www2.ed.gov/about/offices/list/ocr/letters/colleague-20071226.html.

California State University or the University of California system. In 2015, Asian students were about 15 percentage points more likely than white students to have taken the necessary high school coursework and about 30 percentage points more likely than black or Hispanic students (Gao, 2016). Similarly, selective colleges often use the difficulty of the students' coursework as an important measure of student effort in making admissions decisions (Jaschik and Lederman, 2018): students who are placed in lower tracks are therefore at a disadvantage. And since many universities offer college credit for students who take and pass AP and similar exams, students who attend schools without those opportunities may be at a disadvantage.

Advanced course taking in high school is a strong indicator of opportunity to learn because it reflects both systematic differences in the availability of these courses and in who participates in them. As such, improving access to high-quality advanced coursework across several disciplines represents a potential lever for reducing group disparities in educational attainment.[12]

Tables 5-3 through 5-5 show the percentage of schools with high and low populations of black and Latino students ("high-minority" schools, "low-minority" schools)[13] that do not offer higher level math and science courses and AP, International Baccalaureate (IB), and dual enrollment programs. Tables 5-6 through 5-8 show the same information for schools with high and low percentages of economically disadvantaged students ("high-poverty" schools, "low-poverty" schools). In most cases, high-minority schools are at least twice as likely to not offer these courses as are low-minority schools. Similarly, high-poverty schools are at least 1.5 times as likely as low-poverty schools to not offer advanced coursework in math and science and to not offer AP courses or dual enrollment programs.

Proposed Measures for Indicator 11

Indicator 11 should be measured by differential rates of enrollment and participation in gifted and talented programs, the coursework needed for college preparation, AP and IB courses, and dual enrollment programs.

---

[12]See https://nces.ed.gov/programs/digest/d10/tables/dt10_049.asp.

[13]We use the terms "high-minority schools" and "low-minority schools" as they are used by others; "minority" refers to black and Latino students.

TABLE 5-3 Number and Percentage of Schools with No Access to Core Math Courses, by Percentage of Racial/Ethnic Minority Students in Quintiles (Q)

Subject	Quintile 1 Low Minority 20th percentile or lower in percent of minority students		Quintile 2		Quintile 3		Quintile 4		Quintile 5 High Minority 80th percentile or higher in percent of minority students		All Schools
	%	#	%	#	%	#	%	#	%	#	%
Algebra 1 or higher	12%	764	13%	689	17%	939	21%	1,246	25%	1,371	20%
Geometry or higher	13%	853	15%	790	20%	1,097	26%	1,502	29%	1,591	23%
Algebra 2 or higher	15%	980	19%	987	24%	1,307	32%	1,864	35%	1,962	28%
Advanced Math or higher	22%	1,395	29%	1,526	36%	1,988	47%	2,740	51%	2,841	39%
Calculus	40%	2,542	46%	2,446	52%	2,893	62%	3,625	70%	3,887	55%

SOURCE: ExcelinEd (2018, p. 16). Reprinted with permission from Foundation for Excellence in Education.

TABLE 5-4 Number and Percentage of Schools with No Access to Core Science Courses, by Percentage of Racial/Ethnic Minority Students in Quintiles (Q)

Subject	Quintile 1 Low Minority 20th percentile or lower in percent of minority students		Quintile 2		Quintile 3		Quintile 4		Quintile 5 High Minority 80th percentile or higher in percent of minority students		All Schools
	%	#	%	#	%	#	%	#	%	#	%
Biology or higher	14%	877	15%	798	20%	1,122	26%	1,513	29%	1,592	23%
Chemistry or higher	18%	1,163	23%	1,233	30%	1,662	39%	2,310	42%	2,312	33%
Physics	31%	1,990	37%	1,977	44%	2,443	53%	3,111	59%	3,265	47%

SOURCE: ExcelinEd (2018, p. 17). Reprinted with permission from Foundation for Excellence in Education.

TABLE 5-5 Number and Percentage of Schools with No Access to Advanced Placement, International Baccalaureate, or Dual Enrollment Courses, by Percentage of Racial/Ethnic Minority Students in Quintiles (Q)

Program	Quintile 1 Low Minority 20th percentile or lower in percent of minority students		Quintile 2		Quintile 3		Quintile 4		Quintile 5 High Minority 80th percentile or higher in percent of minority students		All Schools
	%	#	%	#	%	#	%	#	%	#	%
Advanced Placement	48%	3,098	48%	2,541	50%	2,769	58%	3,427	62%	3,411	55%
International Baccalaureate	99%	6,313	97%	5,154	96%	5,303	95%	5,594	97%	5,380	97%
Dual Enrollment	33%	2,088	41%	2,160	49%	2,693	61%	3,579	69%	3,820	52%

SOURCE: ExcelinEd (2018, p. 17). Reprinted with permission from Foundation for Excellence in Education.

TABLE 5-6 Number and Percentage of Schools with No Access to Core Math Courses, by Percentage of Economically Disadvantaged Students in Quintiles (Q)

Subject	Quintile 1 Low Poverty 20th percentile or lower in percent of low-income students		Quintile 2		Quintile 3		Quintile 4		Quintile 5 High Poverty 80th percentile or higher in percent of low-income students		All Schools
	%	#	%	#	%	#	%	#	%	#	%
Algebra 1 or higher	16%	838	10%	651	12%	699	21%	1,025	21%	835	20%
Geometry or higher	18%	940	12%	737	14%	811	25%	1,214	26%	1,031	23%
Algebra 2 or higher	21%	1,105	14%	897	18%	1,006	31%	1,523	32%	1,281	28%
Advanced Math or higher	29%	1,528	21%	1,348	29%	1,631	46%	2,274	50%	1,975	39%
Calculus	39%	2,105	38%	2,373	49%	2,750	66%	3,253	72%	2,829	55%

SOURCE: ExcelinEd (2018, p. 18). Reprinted with permission from Foundation for Excellence in Education.

TABLE 5-7 Number and Percentage of Schools with No Access to Core Science Courses, by Percentage of Economically Disadvantaged Students in Quintiles (Q)

Subject	Quintile 1 Low Poverty 20th percentile or lower in percent of low-income students		Quintile 2		Quintile 3		Quintile 4		Quintile 5 High Poverty 80th percentile or higher in percent of low-income students		All Schools
	%	#	%	#	%	#	%	#	%	#	%
Biology or higher	18%	978	12%	776	15%	835	25%	1,216	25%	991	23%
Chemistry or higher	25%	1,345	18%	1,103	22%	1,261	37%	1,843	41%	1,629	33%
Physics	33%	1,787	30%	1,885	38%	2,157	54%	2,670	60%	2,366	47%

SOURCE: ExcelinEd (2018, p. 19). Reprinted with permission from Foundation for Excellence in Education.

TABLE 5-8 Number and Percentage of Schools with No Access to Advanced Placement, International Baccalaureate, or Dual Enrollment Courses, by Percentage of Economically Disadvantaged Students in Quintiles (Q)

Program	Quintile 1 Low Poverty 20th percentile or lower in percent of low-income students		Quintile 2		Quintile 3		Quintile 4		Quintile 5 High Poverty 80th percentile or higher in percent of low-income students		All Schools
	%	#	%	#	%	#	%	#	%	#	%
Advanced Placement	41%	2,172	42%	2,667	50%	2,822	62%	3,042	65%	2,574	55%
International Baccalaureate	96%	5,137	97%	6,070	96%	5,400	97%	4,784	97%	3,847	97%
Dual Enrollment	44%	2,335	34%	2,160	40%	2,266	58%	2,858	65%	2,562	52%

SOURCE: ExcelinEd (2018, p. 19). Reprinted with permission from Foundation for Excellence in Education.

Indicator 12: Disparities in Curricular Breadth

A core mission of America's schools is to produce "active, informed members of a democratic society" (Campbell, 2018, p. 1). A broad curriculum that includes courses in art, geography, history, civics, technology, music, science, world languages, and other subjects can contribute to this mission and help students become well-rounded individuals. While it is not known which specific combination of courses is best for students' long-term outcomes, any educational system that differentially deprives students of exposure to a broad range of subjects is inequitable.

Every state has educational standards for a comprehensive range of subjects that, in theory, contributes to the broad education of all students and fulfills the mission of preparing them to participate in civic society. However, the emphasis that states place on those subjects and the resources they devote varies greatly. In addition, in the No Child Left Behind era, many subjects were eclipsed by an intense national focus on mathematics and reading (Dee, Jacob, and Schwartz, 2013; Ladd, 2017).

Decades of research have demonstrated that schools under the most pressure to improve test scores for purposes of accountability—which are almost always schools serving high proportions of black, Hispanic, and low-income students—often respond by narrowing the curriculum. In these schools and school systems, it is common:

- to focus on tested subjects and, within those subjects, on assignments that mimic standardized tests in terms of content and form (Au, 2007; Darling-Hammond and Wise, 1985; Firestone, Maryowetz, and Fairman, 1998; Madaus, 1988; Meherens, 1998; Newmann, Bryk, and Nagaoka, 2001);
- to restructure the school day to focus more intensively on core content areas (e.g., creating a block for literacy) (U.S. Government Accountability Office, 2009); and
- to spend more time on general test-taking strategies (U.S. Government Accountability Office, 2009).

Although the committee knows of no specific work linking these forms of curricular narrowing to later educational outcomes, this gap is in part due to the available data. For instance, it is not possible to know whether reducing instructional time on social studies has affected students' civic knowledge because there is no comparable measure of civics knowledge to use as a dependent variable in such an analysis. Nevertheless, there is clear conceptual support for the idea that students' access to a broad, diverse curriculum should not be determined by their personal characteristics or the characteristics of the schools they attend—thus, we believe curricular breadth may be an important equity issue that merits further attention. However, we acknowledge that resource constraints will create tradeoffs when seeking to maximize overall equity. Curricular breadth in a high-needs school may be less important than, for example, academic supports for struggling students.

Proposed Measures for Indicator 12

Until measures of curricular breadth are developed and widely used, we encourage their consideration at the local level.

Indicator 13: Disparities in Access to High-Quality Academic Supports

Many students come to school needing resources or supports beyond those that are provided to most students. Some may need services to improve their English proficiency, and others may need special education services to address learning challenges. Still others may not need placement in a formalized program but could benefit from short-term tutoring or other individualized academic supports. The need for school-based academic supports is often greater when schools have a higher concentration of financially disadvantaged students, but the available resources may be less than adequate. School-based academic supports can include a variety of services, such as academic support classes, academic tutoring, early warning systems, and high school transition activities. Research suggests that academic support classes may have a positive effect on such student outcomes as average number of credits earned, high school graduation, and college enrollment.[14] Data on disparities in the prevalence of these services are available through the U.S. Department of Education 2017 National Survey on High School Strategies Designed to Help At-Risk Students Graduate.[15] The data are reported by school characteristic (e.g., high poverty, low poverty) and for groups of students likely to need extra help.

It is crucial that schools provide supports to address students' academic, linguistic, and special education needs. There is a risk, however, that students might be identified for services that they do not need and that could result in reduced access to high-quality, appropriate instruction for those students.

Given the overrepresentation of some racial and ethnic groups in special education programs, educators have to be careful to ensure that students are not misidentified for those services. Inaccurate special education placement can be detrimental to students' educational trajectories, locking stu-

---

[14]See https://www2.ed.gov/rschstat/eval/high-school/academic-tutoring.pdf.
[15]See https://www2.ed.gov/about/offices/list/opepd/ppss/reports-high-school.html.

dents into a less demanding curriculum and potentially limiting exposure to challenging courses and instruction and to a diverse group of learners. The overrepresentation of black and Latino students in special education can also contribute to racially and ethnically segregated classrooms. Reports of this type of segregation have increased over time (see National Council on Disability, 2018)[16] and have been the subject of lawsuits brought by the U.S. Department of Justice.[17] Similarly, isolating English learners can limit their opportunities to participate in rigorous and challenging instructional programs and is likely to reduce opportunities for interactions between these students and their English-proficient peers. Concerns about linguistic isolation have increased over time and also been the subject of lawsuits.[18]

Proposed Measures for Indicator 13

The committee's proposed indicator focuses on access to and participation in formalized systems of tutoring or other types of academic supports. Equity requires that all students have opportunities to participate in these kinds of programs if they demonstrate a clear need for them. At the same time, given the concerns about excessive identification of some students for special services, the indicator would need to address the extent to which services are appropriately matched to a student's needs. To address racial, ethnic, and linguistic disparities in placement into special programs, an indicator system would need to monitor rates of identification and program placement for each group and document the extent to which identification and placement policies are applied in an equitable way. In the case of special education services, we would need measures of rates of identification in various disability categories by group and of the restrictiveness of the placements (e.g., whether and how much time students spend in separate classrooms or schools).

DOMAIN G: EQUITABLE ACCESS TO SUPPORTIVE SCHOOL AND CLASSROOM ENVIRONMENTS

Students need more than challenging courses and effective teachers to thrive academically. They also need physically and emotionally safe learning environments, with a range of supports that pave the way for them to succeed by addressing their socioemotional and academic needs. Safe, sup-

---

[16]See https://ncd.gov/sites/default/files/NCD_Segregation-SWD_508.pdf.

[17]See https://www.washingtonpost.com/news/education/wp/2016/08/23/justice-department-sues-georgia-over-segregation-of-students-with-disabilities/?utm_term=.f5bab3bdb748.

[18]See, for example, https://datacenter.kidscount.org/updates/show/150-linguistic-isolation-still-a-challenge and https://civilrightsproject.ucla.edu/legal-developments/court-decisions/the-educational-implications-of-linguistic-isolation-and-segregation-of-latino-english-language-learners-ells.

portive school environments and access to academic supports, counseling, and referral to social services are especially important for students who experience chronic stressors outside of school that affect their learning and development.

This domain addresses some key school-based features that influence students' opportunities to learn: strong school climate, the use of preventative, nonexclusionary discipline policies, and socioemotional and mental health supports. Although some of these indicators are difficult to define and measure, they are important to include in a system of educational equity indicators because research is establishing a relationship between these factors and student outcomes and because there is evidence to suggest that there are group differences in access to these supportive factors.

Indicator 14: Disparities in School Climate

School climate is increasingly recognized as an important influence on many student outcomes, with evidence that a healthy climate links directly to achievement, graduation rates, and effective risk prevention (Allensworth and Easton, 2007; Cohen and Geier, 2010; Faster and Lopez, 2013; MacNeil, Prater, and Busch, 2009; Thapa et al., 2012; Wang et al., 2014). Definitions of school climate vary widely, but, in general, "climate" refers to the way that a school feels to students, as well as to adults who work in the buildings and to family members (Kostyo, Cardichon, and Darling-Hammond, 2018). Aspects of climate can include safety, supportiveness of staff, absence of harassment and discrimination, connectedness among students and staff, sense of fairness, and trust of adults and other peers, among other factors.

There is some evidence that positive school climate is associated with improved outcomes for students, but moreover, schools with hostile climates can negatively affect at-risk students, having been linked to depression, low self-esteem, feelings of victimization, and lower academic achievement (Kosciw et al., 2012; O'Malley et al., 2014; Thapa et al., 2013). At present, data on equitable access to safe, supportive climates are not widely available. In one statewide analysis of Illinois schools, proportionally fewer Chicago public schools had supportive climates than suburban schools and schools in other urban areas (Klugman et al., 2015). Rural and small-town schools were the least likely to have supportive climates. Even for students in the same class and the same school, perceptions of climate can differ. Differences in perceptions of climate across population groups would help identify differences in access to supportive environments.

Climate can be measured through a variety of approaches, including surveys of students, staff, and family members, structured observations of school and classroom environments, and reviews of documentation on such factors as school conditions and resource availability. Surveys are generally the most suitable method for inclusion in a large-scale data collection effort because they are relatively inexpensive and can be designed to gather perceptions about a broad range of aspects of climate. In addition, survey data can be disaggregated to examine disparities across groups of students.

Although surveys about climate are not routinely administered to schools across the country, several states have adopted climate measures for use in their accountability systems under the Every Student Succeeds Act (ESSA), and many school districts also administer climate surveys.[19]

The Chicago Consortium on School Research has developed robust measures and collected extensive longitudinal data on school climate. Research from Chicago public schools—where nearly 80 percent of students are socioeconomically disadvantaged—as well as the state of Illinois, has shown that students have higher academic achievement in schools in which staff and students report positive school climates than in schools in which staff and students report weak school climates, comparing schools serving students with similar backgrounds (Brookover et al., 1979; Bryk et al., 2010; Haynes, Emmons, and Ben-Avie, 1997; Klugman et al., 2015; Tschannen-Moran, Parish, and DiPaola, 2006). Moreover, improvements in learning gains are higher in schools that have safe, academically focused climates than in those that do not and in schools that see improvements in school climate (Sebastian and Allensworth, 2012; Sebastian, Allensworth, and Stevens, 2014; Sebastian, Huang, and Allensworth, 2017). Other research has found that teacher qualifications were only related to student achievement in schools with safe climates (DeAngelis and Presley, 2011).

Other states and school districts have also developed ways to evaluate school climate:

- Massachusetts: In 2017 the state began collecting information related to students' socioemotional learning, health, and safety through a separate questionnaire in the Massachusetts Comprehensive Assessment System, the state's standardized assessment (Massachusetts Department of Elementary and Secondary Education, 2017).
- California: Every 2 years, California's public school system administers the California Healthy Kids Survey to school staff and students in grades 7-9 to measure climate, student engagement in learning, health, and well-being (Austin et al., 2016).
- Nevada: To address educational equity as required by ESSA, Nevada is developing a school climate and social and emotional learning measure. The survey will be administered to students in grades 5-12

---

[19]See https://learningpolicyinstitute.org/product/essa-equity-promise-climate-brief.

and will serve as a needs assessment to inform future efforts related to school climate (U.S. Department of Education, 2017).

The U.S. Department of Education also created a center to help school systems develop safe and supportive learning environments[20] and compiled climate survey items that are available for public use.[21] The National Institute of Justice recently issued a report focused on creating and sustaining a positive and communal school climate,[22] and the Office of Civil Rights issued a report summarizing its findings from data collections on climate and safety in public schools.[23]

Proposed Measures for Indicator 14

At present, measures of school climate are not ready to be included in a nationwide system of equity indicators. We anticipate that measures of school climate will become more widely used as states work to comply with ESSA. In the meantime, we encourage their use at the local level.

Indicator 15: Disparities in Nonexclusionary Discipline Practices

A school's approach to student discipline can influence students' access to equitable learning conditions. Exclusionary discipline policies, such as in- or out-of-school suspension, remove students from the classroom, thereby reducing their opportunities to learn from instruction. As a result, these practices could negatively affect student learning and other outcomes for students who are subjected to them.

Suspensions are often imposed even for such relatively minor and nonviolent infractions as tardiness or failure to show respect to adults (González, 2012). Research suggests that suspension rates are negatively correlated with student achievement (Skiba et al., 2014) and positively correlated with a student's likelihood of dropping out of school (Flannery, 2015; Rumberger and Losen, 2016). In Chicago, conversely, reductions in the use of suspensions were associated with improvements in students' test scores and attendance and in perceptions of climate in schools with majority black students (Hinze-Pifer and Sartain, 2018). Suspensions themselves are also associated with school climates that are less safe (Steinberg, Allensworth, and Johnson, 2015).

---

[20]See https://safesupportivelearning.ed.gov/.
[21]See https://safesupportivelearning.ed.gov/edscls.
[22]See https://www.ncjrs.gov/pdffiles1/nij/250209.pdf.
[23]See https://www2.ed.gov/about/offices/list/ocr/docs/school-climate-and-safety.pdf.

Addressing suspensions is particularly relevant to equity concerns given the large discrepancies in suspension rates across racial and ethnic groups. In California schools, for example, black students were subject to harsher disciplinary actions, including suspensions, compared with their white counterparts (Losen, Martinez, and Gillespie, 2012). Overall, black students tend to be subjected to harsher disciplinary consequences than white students, even for the same infractions in the same schools (Anderson and Ritter, 2017). More broadly, students from underrepresented groups, including students with disabilities, are suspended at disproportionate rates (U.S. Department of Education, Office for Civil Rights, 2016; U.S. Government Accountability Office, 2018). Evidence suggests that these differences can negatively affect short- and long-term outcomes for some students. Figure 5-2 shows suspension rates nationally for various groups of students.

In recent years, many school districts across the United States have enacted new disciplinary approaches that aim to reduce exclusionary practices. Many of these approaches can be classified as "restorative practices," which aim to help students build high-quality relationships and develop conflict-resolution skills (Advancement Project, 2014; Fronius et al., 2016). Rigorous research on the effects of these changes to disciplinary policy, or on the supports that educators need to implement them effectively, is limited, but the practices do provide one potential avenue for reducing disruptions in learning due to disciplinary events. It is currently not possible to measure schools' use of nonexclusionary disciplinary policies, the extent to which teachers are trained to use nonpunitive approaches, or the extent to which they effectively implement these approaches. More-

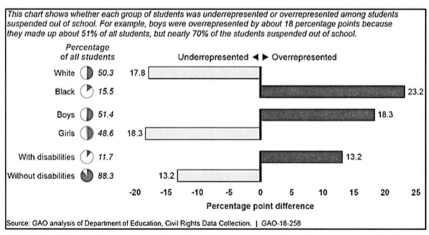

FIGURE 5-2 Student suspensions by race, sex, and disability status, school year 2013-2014, K–12 public schools.
SOURCE: U.S. Government Accountability Office (2018, Highlights).

over, because the research on the effectiveness of these approaches is so limited, we are not endorsing the creation of an indicator of specific non-exclusionary discipline practices. Until such measures can be developed and the research base becomes more solid, suspension and expulsion rates can be used to indicate the absence of nonexclusionary methods. Although suspension rates are problematic as an indicator of responsiveness, disproportionality in suspensions is a much stronger indicator of inequity. Educators and policy makers can use these data to gather more information on the reasons for group differences that are revealed by differences in suspension rates—including differences in school or district policy—and address those causes.

Proposed Measures for Indicator 15

Suspension and expulsion rates are already being reported by states, as required under ESSA, and are ready for inclusion in the system we propose. Both in-school and out-of-school rates should be tracked.

Indicator 16: Disparities in Nonacademic Supports for Student Success

There are many ways that schools can help support students at risk of school failure. We grouped these supports into four categories, as explained below:

- Socioemotional development: examples include using specific curricular programs, embedding socioemotional learning practices into curriculum and instructing, and embedding socioemotional development into the school climate.
- Meeting the emotional, behavioral, and mental health needs of students who are exposed to violence and other stressors in their homes and neighborhoods: examples of such supports might involve providing onsite counseling or appropriate referral services that help students respond to the traumas that they face.
- Physical health: examples include providing dental or medical screenings for students who otherwise may not have access.

Research on the efficacy of these supports and their links to student outcomes is as wide-ranging as the supports themselves. In some areas, such as the effects of additional instructional time on student achievement, the research is well developed and shows that targeted increases in instructional time, such as through double-dose algebra programs, can boost achievement and attainment.[24] In contrast, research on how schools can develop students' socioemotional competencies[25] is less developed. Research suggests that socioemotional competencies are important because they predict later outcomes, including success in the labor market (Deming, 2015; Schanzenbach et al., 2016). Several research syntheses indicate that some instructional programs designed to promote socioemotional skill development have positive effects not only on those skills, but also on a variety of short- and long-term student outcomes, including academic achievement, disciplinary incidents, and postsecondary success (Grant et al., 2017; Taylor et al., 2018; Weissberg et al., 2015). In addition, schools can promote socioemotional development through means other than adopting an explicit socioemotional skills curriculum, including through supportive school climates and the adoption of instructional practices that support the development of student agency, collaboration, and related skills (Allensworth et al., 2018; Jones and Kahn, 2017). However, the research does not support conclusions about which of these approaches might be most effective in any given context, and high-quality assessments of socioemotional competencies that could be incorporated into a large-scale indicator system are limited (see Taylor et al., 2018).

This committee's determination is that schools and districts ought to be providing supports to meet the needs of their populations. Research suggests that the need for extra resources increases commensurate with the rate at which schools serve students with disabilities, English learners, and students from financially disadvantaged families (see Duncombe and Yinger, 2005; Gandara and Rumberger, 2008; Gronberg, Jansen, and Taylor, 2011). This indicator would involve measuring student needs at a school and tracking the level of student supports as it relates to those needs. Educational jurisdictions could track, for example, the provision of before- and after-school programs, free supplemental academic tutoring, the ratio of school counselors and school psychologists to students, and the availability of school nurses. The specific measures of this indicator would be determined locally because they would depend on the needs of the student population in a given educational jurisdiction.

Of course, having such a tailored response requires resources. Although some states have improved the equitable distribution of their resources—that is, students with greater needs receive more resources—there are still states where low-income students receive fewer resources than their more

---

[24]See http://jhr.uwpress.org/content/50/1/108.short and https://www.aeaweb.org/articles?id=10.1257/aer.104.5.400.

[25]Socioemotional competencies include interpersonal skills such as teamwork and social awareness and intrapersonal skills such as self-regulation and persistence (Taylor et al., 2018).

affluent peers. The most recent causal research suggests that equity-oriented finance reforms boost educational outcomes, especially for students at risk of school failure (Jackson, 2018; Jackson, Johnson, Persico, 2016; Lafortune, Rothstein, and Whitmore-Schanzenbach, 2016).

Proposed Measures for Indicator 16

Measures for Indicator 16 should include the availability of supports for socioemotional development; emotional, behavioral, and mental health; and physical health. These measures are not yet available at a national level but should be tracked at a local or state level.

REVIEW OF EXISTING DATA SOURCES AND PUBLICATIONS

A key part of the committee's work was to investigate the potential usefulness of existing data systems and indicator reports for our proposed indicator set. Box 4-2 in Chapter 4 shows the criteria we used. Overall, while there is a wealth of information on pre-K to grade 12 education, the existing data and reports are not sufficient for the set of educational equity indicators as we have conceptualized them. Relevant information is scattered across multiple databases, which define some indicators and measure some constructs in different ways, do not provide any measures for some constructs, vary in data collection procedures, frequency, geographic detail, and coverage of student groups of interest, and are accessible through different agencies and organizations.

Tables 5-9, 5-10, 5-11, and 5-12 summarize the potential data sources for each of the nine indicators and specific constructs for each indicator that we propose for Domains D, E, F, and G, respectively. The tables also summarize the extent to which data are ready with which to develop specific measures of each construct, and if not ready, what is needed. These tables draw on the information on existing data systems in Appendix A, existing publications that include indicators of educational equity in Appendix B, and our assessment of data and methodological challenges and opportunities for educational equity indicators in Appendix C.

For these domains and indicators, the constructs and measures used pertain to students, categorized by groups of interest, in schools with specific characteristics. At the school level, a measure could be whether the student body is predominantly low, middle, or high income, for example—defining those three categories relative to the district as a whole, the state, or the nation. Corresponding measures for multi-school districts, states, and the nation would be the percentage of students in each group of interest attending schools with student bodies that are predominantly low, middle, or high income, however defined (see Appendix C). As noted

TABLE 5-9 Potential Data Sources and Measures for Domain D, Extent of Racial, Ethnic, and Economic Segregation

Constructs	Source (Characteristics)
Indicator 8: Disparities in Students' Exposure to Racial, Ethnic, and Economic Segregation	
Concentration of Poverty in Schools	**Source:** ED*Facts* (as part of ESSA reporting requirements) **Frequency:** Annual
Racial Segregation Within and Across Schools	**Geographic detail:** Nation, states, districts, schools (elementary, middle, secondary, other) **Student group detail:** Race/ethnicity, gender, English-language status, disability status, economically disadvantaged (typically eligible/not eligible for free or reduced-price lunch) **Possible measures:** Percent students attending low-income, middle-income, high-income schools; percent students attending schools with high, medium, and low percentages of specified race/ethnicity groups **Future potential:** In the case of poverty, because states generally use NSLP eligibility as their indicator of low income, work is needed to develop an appropriate measure—e.g., by having the Census Bureau model ACS poverty data for school attendance areas or student bodies (see Appendix C)

NOTES: ACS, American Community Survey; NSLP, National School Lunch Program.

in Chapter 4, an appropriate measure of poverty and income status more generally is needed to replace the less and less appropriate measure commonly used—namely, eligibility for free or reduced-price school lunch (see Appendix C). Note also that the Civil Rights Data Collection (CRDC) is a lead data source for many constructs in these four domains.

TABLE 5-10 Potential Data Sources and Measures for Domain E, Equitable Access to High-Quality Early Learning Programs

Constructs	Source (Characteristics)
Indicator 9: Disparities in Access to and Participation in High-Quality Pre-K Programs	
Availability of and Participation in Licensed Pre-K Programs	**Source:** CRDC
	Frequency: Biannual
	Geographic detail: Nation, states, districts
	Student group detail: Race/ethnicity, gender, English-language status, disability status
	Possible measures: Percent students in districts that offer pre-K; percent children ages 3-5 enrolled in pre-K when offered by district
	Comment: This is a proxy measure that does not address quality of programs; also, the CRDC does not capture information on other licensed programs such as Head Start (NIEER surveys of states asks about all pre-K programs they fund plus Head Start and special education—see Appendix B)
	Future potential: Substantial work would be required to develop standard rating systems for pre-K quality for programs offered by districts and other organizations, building on state experience with various rating systems; NIEER's surveys ask basic facts that could contribute to a quality measure, including hours pre-K offered, whether teachers have a B.A., teacher-student ratio, etc.

NOTES: CRDC, Civil Rights Data Collection; NIEER, National Institute for Early Education Research.

TABLE 5-11 Potential Data Sources and Measures for Domain F, Equitable Access to High-Quality Curricula and Instruction

Constructs	Source (Characteristics)
Indicator 10: Disparities in Access to Effective Teaching	
Teachers' Years of Experience	**Source (1): NTPS**
	Frequency: Biannual
	Geographic detail: Nation (sample too small for finer detail)
	Student group detail: Nothing for students, but has school level (elementary, middle, secondary) and percent students eligible for free or reduced-price lunch
	Possible measures: Percent all students (nationwide) attending schools (by level and percent NSLP) with low, moderate, high percent teachers with, say, 5+ years teaching (select threshold based on evidence of teaching effectiveness)
	Source (2): CRDC
	Frequency: Biannual
	Geographic detail: Nation, states, districts, schools (elementary, middle, secondary school, other)
	Student group detail: Race/ethnicity, gender, English-language status, disability status
	Possible measures: Percent students attending schools with low, moderate, high percent teachers with 2+ years teaching
	Future potential: The CRCD only distinguishes teachers with 1, 2, or 3+ years teaching; could construct from SLDS as more states develop them in a comparable manner, include information on teacher experience, and provide access for statistical purposes

continued

TABLE 5-11 Continued

Constructs	Source (Characteristics)
Teachers' Credentials, Certification	**Source:** CRDC
	Frequency: Biannual
	Geographic detail: Nation, states, districts, schools (elementary, middle, secondary, other)
	Student group detail: Race/ethnicity, gender, English-language status, disability status
	Possible measures: Percent students attending schools with low, moderate, high percent fully certified teachers; percent students in grades 7-8 and 9-12 attending schools with all math classes taught by teachers certified in math
Racial and Ethnic Diversity of the Teaching Force	**Source:** NTPS (see Teachers' Years of Experience, Source (1), above)
	Future potential: Construct from SLDS as more states develop them in a comparable manner, include information on race/ethnicity of teachers, and provide access for statistical purposes; could possibly construct measures of percent students in schools with high, medium, or low percent teachers with the same race/ethnicity as majority of students in school and classroom

Indicator 11: Disparities in Access to and Enrollment in Rigorous Coursework

Constructs	Source (Characteristics)
Availability and Enrollment in Advanced, Rigorous Coursework	**Source:** CRDC
	Frequency: Biannual
	Geographic detail: Nation, states, districts, middle schools
	Student group detail: Race/ethnicity, gender, English-language status, disability status
	Possible measures: Percent students in grades 7-8 in middle schools that offer algebra I; percent students in grade 8 enrolled when offered (also has information on high school enrollment in various science and math courses)
	Future potential: Construct more complete measures from transcript information from SLDS as more states develop them in a comparable manner and provide access for statistical purposes

TABLE 5-11 Continued

Constructs	Source (Characteristics)
Availability and Enrollment in AP, IB, Dual Enrollment, and Gifted and Talented Programs	**Source:** CRDC **Frequency:** Biannual **Geographic detail:** Nation, states, districts, high schools **Student group detail:** Race/ethnicity, gender, English-language status, disability status **Possible measures:** Percent students in high schools that offer AP (IB) (DE) (G&T) courses; percent students enrolled when offered (also has information on G&T in elementary and middle schools)

Indicator 12: Disparities in Curricular Breadth

Availability and Enrollment in Coursework in the Arts, Social Sciences, Sciences, and Technology	**Source:** None at present (except that CRDC has information on science and computer science classes in high schools) **Desirable possible measures:** Percent students in schools (by level) that offer complete range of subjects by (most generous) state standards **Future potential:** Construct from SLDS as more states develop them in a comparable manner and provide access for statistical purposes

Indicator 13: Disparities in Access to High-Quality Academic Supports

Access to and Participation in Formalized Systems of Tutoring or Other Types of Academic Supports	**Source:** None at present (except that CRDC has information on numbers of FTE instructional aides; it also has information about access to and enrollment in various courses for student groups, which could help identify the extent to which English-language learners and students with disabilities are receiving appropriate academic support) **Desirable possible measures:** Percent students in schools (by level) that have low, medium, high ratios of tutors, counselors, and other support staff per student; percent students using such resources **Future potential:** Further research and data collection are needed to develop useful measures for this construct

NOTES: AP, Advanced Placement; CRDC, Civil Rights Data Collection; DE, dual enrollment (high school and college); FTE, full-time equivalent; G&T, gifted and talented; IB, International Baccalaureate; NTPS, National Teacher and Principal Survey; SLDS, Statewide Longitudinal Data System.

TABLE 5-12 Potential Data Sources and Measures for Domain G, Equitable Access to Supportive School and Classroom Environments

Constructs	Source (Characteristics)
Indicator 14: Disparities in School Climate	
Perceptions of Safety, Academic Support, Academically Focused Culture, and Teacher-Student Trust	**Source (1):** NCES 2012 EDSCS Pilot, School Safety and Environment Modules
	Frequency: One-time; survey instruments are provided to states, school districts, and schools for their use
	Geographic detail: Nation (sample too small for finer detail)
	Student group detail: Race/ethnicity, gender, whether in special education
	Grade/level detail: Tested questions are available for school climate, including safety and environment (physical, instructional), as seen by 5th- to 12th-grade students, staff, and parents
	Possible measures: Percent students in schools (grades 5-6, 7-8, 9-12) with scale scores above a specified level
	Future potential: Use tested questions in EDSCS to develop assessments that are feasible to administer by schools at scale, nationwide and annually
	Source (2): CRDC
	Comment: School administrators provide counts on harassment, bullying, and school safety, which could be aggregated into one or more scales; however, these include only those incidents known to administrators

TABLE 5-12 Continued

Constructs	Source (Characteristics)

Indicator 15: Disparities in Nonexclusionary Discipline Practices

Out-of-School Suspensions and Expulsions	**Source:** CRDC

Frequency: Biannual

Geographic detail: Nation, states, districts, schools

Student group detail: Race/ethnicity, gender, English-language status, disability status

Possible measures: Percent students in schools with low, moderate, high percentages of suspended/expelled students; percent students in specified group (e.g., male) suspended/expelled at rates above, about the same, and below the average of all students for their school, district, state, and nation

Comment: At present, there are no measures of schools' use of non-exclusionary disciplinary policies; hence, suspension/expulsion rates are a proxy at one extreme of the discipline continuum

Indicator 16: Disparities in Nonacademic Supports for Student Success

Supports for Emotional, Behavioral, Mental, and Physical Health	**Source:** Could be estimated from CRDC data on costs for FTE staff supporting students (e.g., counselors, health professionals, social workers), together with estimated staff costs needed per students who are nonpoor, English fluent, or nondisabled and per students who are poor, English-language learning, or disabled (see Appendix C)

Frequency: Biannual

Geographic detail: Nation, states, districts, schools

Student group detail: Race/ethnicity, gender, English-language status, disability status

Possible measures: Percent students in schools with more than adequate staff resources for their student bodies, adequate resources, or less than adequate resources

Comment: This is only one way of potentially measuring adequacy of school resources to support the needs of students in terms of emotional and behavioral development and physical and mental health

NOTES: CRDC, Civil Rights Data Collection; EDSCS, ED School Climate Surveys; FTE, full-time equivalent; NCES, National Center for Education Statistics.

6

Paths Forward: Recommendations

In Chapters 1 through 5 of this report we present our rationale and vision for monitoring equity in K–12 education with a set of indicators and more specific measures, which are detailed in Chapters 4 and 5. The committee's intent is for these indicators to form the core of a national program to monitor educational equity by assembling information that can be reported at the national, state, and local levels and can be disaggregated for important population groups. This targeted set of indicators will shed light on the nature of between-group differences in academic achievement and educational attainment and disparities in access to critical educational resources. In so doing, they will provide a scientific basis for policies to address those inequities. This final chapter focuses on implementation: namely, how to transition from identifying research-based indicators of educational equity in concept to implementation of a functioning system.

READINESS OF INDICATORS FOR OPERATIONAL USE

As detailed in Chapters 4 and 5, we propose indicators for seven domains. The first three domains (A, B, C) reflect transition points in students' lives across K–12 education: readiness for the transition into kindergarten, steady progress through the grades, and readiness for the transition to postsecondary endeavors (see Figure 2-2 in Chapter 2). The other four domains (D, E, F, G) reflect structures and resources in the K–12 education system that can mitigate or exacerbate disparities: exposure to racial or socioeconomic segregation, access to high-quality early childhood education, access to high-quality curricula and instruction across all achievement levels, and access to supportive schools. For each of these domains, we propose one or more indicators that reflect factors that are (1) critical to academic success and education attainment and (2) sources of between-group differences. For each indicator, we suggest, mindful of parsimony, constructs to measure, track, and compare between-group differences.

Some of the indicators we propose are ready to implement operationally: measures of the constructs have been developed and subgroup data are available at the school, district, state, and national levels. Attendance (or its converse, absenteeism) is an example: states and districts now have administrative data systems that enable reliable, valid, and accurate calculation of attendance rates (or absentee rates) and that support cross-state comparisons.

Other indicators will require additional work before they can be implemented. For example, while many school systems evaluate readiness skills when children enter kindergarten, the assessments they use differ. They differ in terms of the skills assessed, the methods by which they are assessed, the timing of the assessments, and the way of determining readiness. These assessments can be useful in monitoring between-group differences in readiness at the local level, but further work is needed to implement them on a national basis.

In our view, an equity indicator system must evolve, with structures, processes, and resources for continuous improvement in light of research, experience, and richer consensus. A corollary is that indicators or measures will not initially be identical in sophistication, or perhaps ever. As a practical matter, there will be differences in, among other characteristics, strength of research base; precise measures and data definitions across jurisdictions; availability of data disaggregation by population groups; and complexity arising from independent policy choices when there are no federal standards. For example, except for the federally constructed and administered National Assessment of Educational Progress (NAEP), comparisons of achievement across jurisdictions are difficult because states use different tests for their assessments. For another example, for early childhood education states have different standards for quality, even for licensed providers. Differences in financial accounting systems will complicate comparison of resources. Therefore, an indicator system close to a researcher's *ideal* is impossible, although a *useful* system is possible—especially if designed for continuous improvement, as we discuss below.

Several of our proposed indicators and measures either exist or can be readily constructed from existing data collection systems mandated by federal or state policies; typically, these have the force of law by statute or regulation. Chapters 2, 4, and 5—with additional details in Appendixes A, B, and C—described a vast array of nationwide data elements compiled by the National Center for Education Statistics, other agencies in the U.S.

Department of Education, and the U.S. Census Bureau. Many, though not all, of the committee's proposed indicators of educational equity are available from existing data collection programs. Not all data sources are collected on an annual basis, however, and data sets vary in the extent to which indicators can be disaggregated by population group, in the unit of analysis (state, district, school, classroom), and in the availability of the data after collection. Many of the challenging key constructs (e.g., measures of socioeconomic status) are not always consistent among sources, and some indicators require modeling and linking of more than one data set.

There has been innovative work to develop indicators in specific domains. One example is the work of the Stanford Education Data Archive in using calibration procedures to put state assessment results on a common scale based on the National Assessment of Educational Progress (see description of the Stanford Education Data Archive in Appendix A). Another example is adjusted-per-pupil spending to measure equitable distribution of resources relative to student need (see description of the Education Law Center and Rutgers University collaboration in Appendix A). Such work indicates the potential for analytically meaningful indicators of educational equity, although the effort required for the two examples cited is considerable and, at present, results in a lag in availability.

There are many publications of key indicators for K–12 education and for child well-being more generally, and most publications link to more detailed underlying data. But none of the publications, including those that focus specifically on between-group disparities, presents a fully developed representation based on a carefully articulated concept of equity that covers all student groups of interest. In addition, the indicators in some reports are based on data sources that cannot support subnational detail, although in many cases the indicators could be developed from the bottom up by states and school districts for individual schools if there was a federal or national (meaning multistate) mechanism for coordination. In addition, data collection is not only a technical issue, but also a matter of public policy so there are political considerations and disagreement over such issues as compliance burdens and federalism: Which level of government has the "right" to impose data collection requirements and constructs? For example, the Every Student Succeeds Act (ESSA) contains bare-bones requirements for data collection but leaves many design and implementation choices to states. ESSA is in some respects a federal framework, but it deemphasizes national uniformity by leaving many important choices to individual states. Relatedly, and relevant to the equity discussion, there is now serious political division over the future scope of the Civil Rights Data Collection program.

CONCLUSION 6-1: Existing publications are mixed in their ability to support the committee's proposed set of K–12 educational equity indicators. There is at present limited support for creating within the federal government an expanded structure of nationally uniform data collection.

IMPLEMENTATION

Clearly, refining and implementing the committee's proposed system would be easier if there were available data and indicators that simply needed to be brought together or modified slightly to satisfy the goals of an equity indicator system. Our review reveals that this is not the case. Considerable effort will be needed to assemble the necessary data and conduct the necessary analyses and data manipulations to generate comparable indicators for the nation, states, districts, and schools. Substantial effort will also be required to implement, evaluate, and improve a system of educational equity indicators on a continuing, regular basis.

Yet the committee concludes that such a system, including its continuous improvement, is essential for myriad social, economic, and fairness reasons. As we state throughout this report, a problem cannot be addressed if it is invisible. A system of equity indicators with its design based on research and consensus would, we believe, be a major advance in the nation's capacity to understand and address ubiquitous disparities in education opportunities and outcomes. This is important for each level of government. Indeed, we believe it is an essential step.

Bottom-Up and Top-Down Approaches

We do not anticipate that an indicator system will be built in a day. Rather, we envision that a system would be developed gradually, making use of existing data first, adapting them to the needs of the indicator system, and making use of that beginning until a fuller system is developed. We see two paths forward.

The first path (bottom up) is for a set of "early adopter" states and districts to partner with each other, researchers, stakeholders, and governmental or philanthropic funders to develop the design and implementation details for the indicator system we describe. This effort would provide the early adopters with useful information, but the more important goal would be creation of a prototype for other jurisdictions to adopt. The prototype would build on existing data and indicator systems used by the early adopters and would also be informed by broader analysis of successes and failures in other jurisdictions. This approach would need to grapple with the existing variation in many indicators, both within and across jurisdictions, and offer a way to consolidate indicators or establish comparability with disparate indicators.

The second path (top down) is for the federal government, working with relevant educational intermediaries—particularly national organizations, such as the National Governors Association, the Council of Chief State School Officers, and the Council of the Great City Schools—to develop an initial version of the indicator system we propose. (This is somewhat analogous to development of the Common Core State Standards.) This path would capitalize on data currently available as a first step and would outline a plan and timetable to add other indicators as conceptual, methodological, and data issues are satisfactorily addressed.

We see these paths as complementary rather than mutually exclusive. In the best case, they would operate concurrently and interact so that the on-the-ground experience of the prototypes can inform the national effort, and the national effort can facilitate ways for districts to improve their data and measures.

Oversight and Guidance

This country has a great deal of experience with national indicator systems. In the context of education, NAEP is now more than 40 years old and commands respect throughout the field of education. It is overseen by a governing board (the National Assessment Governing Board [NAGB]) that is charged with its oversight and updates and modifies it so that it continues to be useful.

NAEP provides a model that may be useful in designing the indicators system we propose. We note the involvement of federal, state, and local stakeholders, research firms with survey and test development expertise, and private foundations in the development, evaluation, and improvement of NAEP. We also note the time it took to establish NAEP as a valued national resource, which culminated with congressional action in 1988 to establish the National Assessment Governing Board (NAGB) as the NAEP oversight body and in 2001 to require states to participate in math and reading assessments. NAEP is still evolving, having, for example, recently added a technology and engineering literacy assessment: see Box 6-1 for a chronology of key milestones in NAEP's development.

We see a need for such collaboration in the development of a national equity indicator system and suggest that interested stakeholders use it as a model. In particular, we see value in a governing body that would serve a role analogous to NAGB's role in NAEP and that partners with the National Center for Education Statistics. We also note that other data collection efforts, such as the Current Population Survey of the U.S. Census Bureau, have oversight or advisory boards, as do several international edu-

BOX 6-1
Milestones in the Evolution of the
National Assessment of Educational Progress (NAEP)

Early 1960s: The idea of a national assessment gains impetus, fueled in part by the 1957 Sputnik launch, which led to the 1958 National Education Defense Act; the Civil Rights Act of 1964, which required a report on equality of education opportunity; the successes and challenges of one-time assessments, including Project Talent and the Equality of Educational Opportunity Survey.

1964: NAEP planning begins, with a grant from the Carnegie Corporation to set up the Exploratory Committee for the Assessment of Progress in Education.

1969: The first NAEP assessments are held, covering citizenship, science, and writing performance of national samples of 9-, 13-, and 17-year-old in-school students, as well as out-of-school 17-year-olds. RTI International conducts the assessments under contract to the Education Commission of the States, which has a grant from the National Center for Education Statistics.

1971: Sample sizes for assessments are expanded to represent Hispanic students.

1971-1973: NAEP assesses 9-, 13-, and 17-year-olds every 4 years in reading and math for "Long-Term Trend NAEP," in which content is kept as comparable as possible over time.

1979: The sample of out-of-school 17-year-olds is dropped because of costs.

1982: A review of NAEP requested by NAEP's-then governing body, the Assessment Policy Committee, and funded by the Carnegie Corporation, Ford Foundation, and Spencer Foundation (see Wirtz and Lapointe, 1982), commends NAEP's quality but calls for inclusion of a broader range of stakeholders in the development and interpretation of NAEP test items, for better reporting of results, and for providing more testing services to states and school districts.

1983: The Educational Testing Service (now, ETS) and Westat win a 5-year grant for NAEP with new design features, including matrix sampling to reduce the burden on students, the use of item response theory for summarizing results on a common scale for students given different test booklets, provision for

cation endeavors, such as the Program for International Student Assessment (PISA), the Progress in International Reading Literacy Study (PIRLS), and the Trends in International Mathematics and Science Study (TIMSS).

There are many technical issues to address in formulating the system so the indicators are valid and useful, and a panel of experts to advise on these issues would be essential. There is also a need for input and buy-in from a range of stakeholders who can address not only the content of and process for producing an informative and coherent set of education equity indicators, but also guidance on how users can interpret the results.

testing special needs students, and identification of anchor points on the scale scores to present results.

1987: A second major evaluation of NAEP (see Alexander, James, and Glaser, 1987) recommends a change in the NAEP governing structure.

1988: Congress passes amendments to the Elementary and Secondary Education Act, authorizing an independent governing board, the National Assessment Governing Board (NAGB), to set NAEP policy, prepare assessment frameworks, and execute the initial release of each round of assessments in "The Nation's Report Card."

1990: Voluntary assessments for states begin on a trial basis and become a permanent feature of NAEP; achievement levels (basic, proficient, advanced), each of which indicates what a student is expected to know, replace scale anchoring points.

1990: Congress revises the Education for All Handicapped Children Act and renames it the Individuals with Disabilities Education Act; NAEP includes students with disabilities and with limited English proficiency.

1990-1992: For math and reading, "main NAEP," in which content is modified about every 10 years to reflect changes in school curricula, assesses 4th-, 8th-, and 12th-grade students.

2001: Congress mandates that all states participate in main NAEP every 2 years for 4th- and 8th-grade students and every 4 years for 12th-grade students.

2002: Selected urban districts participate in state-level assessments on a trial basis and continue as the Trial Urban District Assessment.

2009: NAEP science assessment framework is revised and includes interactive computer tasks for a sample of students.

2011: Writing assessment for 8th- and 12th-grade students is administered entirely on computer.

2014: A new technology and engineering literacy assessment is conducted entirely on computer.

SOURCE: Adapted from Beaton et al. (2011).

RECOMMENDATIONS

RECOMMENDATION 1: The federal government should coordinate with states, school districts, and educational intermediaries to incorporate the committee's proposed 16 indicators of educational equity into their relevant data collection and reporting activities, strategic priorities, and plans to meet the equity aspects of the Every Student Succeeds Act.

RECOMMENDATION 2: To ensure nationwide coverage and comparability, the federal government should work with states, school districts, and educational intermediaries to develop a national system of educational equity indicators. Such a system should be the source of regular reports on the indicators and bring visibility to the long-standing disparities in educational outcomes in the United States and should highlight both where progress is being made and where more progress is needed.

RECOMMENDATION 3: In designing the recommended indicator system, the federal government in coordination with states, school districts, and educational intermediaries, should take care that the system enables reporting of indicators for historically disadvantaged groups of students and for specific combinations of demographic characteristics, such as race and ethnicity by gender. The system also should have the characteristics of effective systems of educational equity indicators identified by the committee.

We note that the system of indicators we propose focuses on the role the education system should play in addressing academic disparities. Although unaddressed in this report, other child-serving agencies play an equally important role in helping children. The effects of adversity on a child or adolescent depends not only on individual resilience and natural variations in child development, but also on the child's opportunity for experiences, interventions, and supports that may mitigate or even undo the effects of adversity, both material and psychological. That is, learning obstacles born of context are not student deficits barring success, but student needs that can be met with appropriate opportunities. Research is needed to increase understanding of how various interventions or opportunities map onto individual student needs that are rooted in context. Consensus-building is needed to create indicators and measures that should be included, eventually, in a broader equity indicator system. For many student needs, screening and responses can best be provided outside of the school settings. Therefore, an indicator system that encompasses all the domains of opportunity important for equity would need to monitor how well student success is supported by other child-serving agencies and nonprofit organizations.

RECOMMENDATION 4: Governmental and philanthropic funders should work with researchers to develop indicators of the existence and effectiveness of systems of cross-agency integrated services that address context-related impediments to student success, such as trauma and chronic stress created by adversity. The indicators and measures should encompass screening, intervention, and supports delivered not only by school systems, but also by other child-serving agencies.

A concerted effort is needed to create the system of equity indicators. Demonstration projects and early prototypes will help catalyze interest in the system and test its feasibility and usefulness.

RECOMMENDATION 5: Public and private funders should support detailed design and implementation work for a comprehensive set of equity indicators, including an operational prototype. This work should involve: (1) self-selected "early adopter" states and districts; (2) intermediaries, such as the Council of the Great City Schools, the Council of Chief State School Officers, and the National Governors Association; (3) stakeholder representatives; and (4) researchers. This work should focus on cataloguing the available data sources, determining areas of overlap and gaps, and seeking consensus on appropriate paths forward toward expanding the indicator system to a broader set of states and districts.

Regardless of the path chosen, a system of equity indicators needs input and buy-in from a range of stakeholders. This input is needed to develop a process for producing an informative and coherent set of educational equity indicators, determine the content of the indicators, and ensure that the results will be understood by users. For these purposes, we believe a governing body is needed to provide governance and implementation. We suggest that one analogous to NAGB, which partners with the National Center for Education Statistics for NAEP, could be a useful model.

RECOMMENDATION 6: Public or private funders, or both, should establish an independent entity to govern the committee's proposed educational equity indicators. The responsibilities of this entity would include establishing and maintaining a system of research, evaluation, and development to drive continuous improvement in the indicators, measures of them, reporting and dissemination of results, and the system generally. This entity might be structured like the National Assessment Governing Board and might report on both levels of the various outcomes the committee proposes and equity gaps in those indicators, as the Governing Board currently does with NAEP.

Acting on these recommendations will keep in the public eye a critical goal for the nation: to ensure that all students receive the supports they need to obtain a high-quality pre-K to grade 12 education. Educating all students is fundamental to the nation's ability to grow and develop and to afford all of its people the opportunity to live full and rewarding lives.

References and Bibliography

Aaronson, D., Barrow, L., and Sander, W. (2007). Teachers and student achievement in the Chicago public high schools. *Journal of Labor Economics, 25*(1), 95-135.

Abedi, J., and Herman, J. (2010). Assessing English language learners' opportunity to learn mathematics: Issues and limitations. *Teachers College Record, 112*(3), 723-746.

Achieve, Inc. (2018). *On Track or Falling Behind? How States Include Measures of 9th Grade Performance in Their ESSA Plans.* Washington, DC: Author.

Adelman, H.S., and Taylor, L. (1997). Addressing barriers to learning: Beyond school-linked services and full-service schools. *American Journal of Orthopsychiatry, 67*(3), 408-421.

Adler, N.E., Boyce, T., Chesney, M.A., Folkman, S., and Syme, S.L. (1993). Socioeconomic inequalities in health: No easy solution. *Journal of the American Medical Association, 269*(24), 3140-3145.

Advancement Project. (2014). *Restorative Practices: Fostering Healthy Relationships & Promoting Positive Discipline in Schools.* Washington, DC: Author. Available: https://advancementproject.org/resources/restorative-practices-fostering-healthy-relationships-promoting-positive-discipline-in-schools/ [August 2019].

Aizer, A., Currie, J., Simon, P., and Vivier, P. (2016). *Do Low Levels of Blood Lead Reduce Children's Future Test Scores?* (No. w22558). Cambridge, MA: National Bureau of Economic Research.

Allensworth, E., and Clark, K. (2018). *Are GPAs an Inconsistent Measure of Achievement across High Schools? Examining Assumptions about Grades versus Standardized Test Scores.* UChicago Consortium on School Research. Available: https://consortium.uchicago.edu/publications/are-gpas-inconsistent-measure-achievement-across-high-schools-examining-assumptions [March 2019].

Allensworth, E., and Easton, J.Q. (2005). *The On-Track Indicator as a Predictor of High School Graduation.* UChicago Consortium on School Research. Available: https://consortium.uchicago.edu/publications/track-indicator-predictor-high-school-graduation [January 2019].

Allensworth, E., and Easton, J.Q. (2007). *What Matters for Staying On-Track and Graduating in Chicago Public High Schools: A Close Look at Course Grades, Failures, and Attendance in Freshman Year.* UChicago Consortium on Chicago School Research. Available: https://consortium.uchicago.edu/publications/what-matters-staying-track-and-graduating-chicago-public-schools [Mary 2019].

Allensworth, E., Farrington, C.A., Gordon, M.F., Johnson, D.W., Klein, K., McDaniel, B., and Nagaoka, J. (2018). *Supporting Social, Emotional, & Academic Development: Research Implications for Educators.* UChicago Consortium on School Research. Available: https://consortium.uchicago.edu/sites/default/files/publications/Supporting%20Social%20 Emotional-Oct2018-Consortium.pdf [December 2018].

Allensworth, E., Gwynne, J.A., Moore, P., and de la Torre, M. (2014). *Looking Forward to High School and College: Middle Grade Indicators of Readiness in Chicago Public Schools.* UChicago Consortium on School Research. Available: https://consortium.uchicago.edu/ publications/looking-forward-high-school-and-college-middle-grade-indicators-readiness-chicago [May 2019].

Allensworth, E., Nagaoka, J., and Johnson, D.W. (2018). *High School Graduation and College Readiness Indicator Systems: What We Know, What We Need To Know.* UChicago Consortium on School Research. Available: https://consortium.uchicago.edu/publications/ high-school-graduation-and-college-readiness-indicator-systems-what-we-know-what-we [May 2019].

American Academy of Pediatrics. (2016). Poverty and child health in the United States. *Pediatrics, 137*(4), 1-16.

American Educational Research Association. (2015). AERA statement on use of value-added models (VAM) for the evaluation of educators and educator preparation programs. *Educational Researcher 44*(8), pp. 448-452. Available: https://journals.sagepub.com/ doi/10.3102/0013189X15618385 [August 2019].

American Statistical Association. (2014). *ASA Statement on Using Value-Added Models for Educational Assessment.* Alexandria, VA: Author. Available: https://www.amstat.org/asa/ files/pdfs/POL-ASAVAM-Statement.pdf [April 2019].

Anderson, K.P., and Ritter, G.W. (2017). Disparate use of exclusionary discipline: Evidence on inequities in school discipline from a U.S. state. *Education Policy Analysis Archives, 25*(49).

Anderson, M.J., Citro, C.F., and Salvo, J.J. (Eds.). (2012). *Encyclopedia of the U.S. Census— From the Constitution to the American Community Survey (ACS),* 2nd Edition. Washington, DC: Congressional Quarterly Press.

The Annie E. Casey Foundation. (2010). *2010 Kids Count Data Book: State Profiles of Child Wellbeing.* Baltimore, MD: Author. Available: https://www.aecf.org/resources/the-2010-kids-count-data-book [January 2019].

The Annie E. Casey Foundation. (2018). *2018 Kids Count Data Book: State Trends in Child Well-Being.* Baltimore, MD: Author. Available: https://www.aecf.org/m/resourcedoc/aecf-2018kidscountdatabook-2018.pdf [April 2019].

Ansari, A., and Purtell, K.M. (2017). Activity settings in full-day kindergarten classrooms and children's early learning. *Early Childhood Research Quarterly, 38,* 23-32. Available: http://dx.doi.org/10.1016/j.ecresq.2016.09.003 [January 2019].

Appleton, J.J., Christenson, S.L., and Furlong, M.J. (2008). Student engagement with school: Critical conceptual and methodological issues of the construct. *Psychology in the Schools, 45*(5), 369-386. Available: https://eric.ed.gov/?id=EJ790338 [January 2019].

Au, W. (2007). High-stakes testing and curricular control: A qualitative metasynthesis. *Educational Researcher, 36*(5), 258-267.

Aucejo, E., and Romano, T.F. (2016). Assessing the effect of school days and absences on test score performance. *Economics of Education Review, 55*, 70-87. Available: https://doi.org/10.1016/j.econedurev.2016.08.007 [March 2019].

Austin, G. (2013). The California healthy kids survey: The case for continuation. *Prevention Tactics, 9*(8), 1-7.

Austin, G., Polik, J., Hanson, T., and Zheng, C. (2016). *School Climate, Substance Use, and Student Well-Being in California*, 2013-2015. Results of the fifteenth Biennial Statewide Student Survey, Grades 7, 9, and 11. San Francisco, CA: WestEd Health & Human Development Program.

Baker, B.D., Farrie, D., and Sciarra, D. (2018). *Is School Funding Fair? A National Report Card*, 7th Edition. Newark, NJ: Education Law Center. Available: http://www.edlawcenter.org/assets/files/pdfs/publications/Is_School_Funding_Fair_7th_Editi.pdf [April 2019].

Balfanz, R., and Byrnes, V. (2006). Closing the mathematics achievement gap in high-poverty middle schools: Enablers and constraints. *Journal of Education for Students Placed at Risk, 11*(2), 143-159. Available: http://dx.doi.org/10.1207/s15327671espr1102_2 [January 2019].

Balfanz, R., and Byrnes, V. (2012a). *Chronic Absenteeism: Summarizing What We Know From Nationally Available Data*. Baltimore, MD: Johns Hopkins University Center for Social Organization of Schools.

Balfanz, R., and Byrnes, V. (2012b). *The Importance of Being There: A Report on Absenteeism in the Nation's Public Schools*. Baltimore, MD: Johns Hopkins University School of Education, Everyone Graduates Center.

Balfanz, R., Byrnes, V., and Fox, J. (2015). Sent home and put off track: The antecedents, disproportionalities, and consequences of being suspended in the 9th grade. In D.J. Losen (Ed.), *Closing the School Discipline Gap: Equitable Remedies for Excessive Exclusion*. New York, NY: Teachers College Press.

Balfanz, R., Herzog, L., and MacIver, D.J. (2007). Preventing student disengagement and keeping students on the graduation path in urban middle-grades schools: Early identification and effective interventions. *Educational Psychologist, 42*(4), 223-235.

Baltimore Education Research Consortium. (2011). *Destination Graduation: Sixth Grade Early Warning Indicators for Baltimore City Schools: Their Prevalence and Impact*. Baltimore, MD: Author. Available: http://baltimore-berc.org/pdfs/SixthGradeEWIFullReport.pdf [January 2019].

Barton, P.E., and Coley, R.J. (2007). *The Family: America's Smallest School*. Princeton, NJ: Educational Testing Service.

Bassok, D., Finch, J., Lee, R., Reardon, S., and Waldfogel, J. (2016). Socioeconomic gaps in early childhood experiences: 1998 to 2010. *AERA Open, 2*(3), 1-22.

Baumeister, R.F., and Vohs, K.D. (Eds.). (2004). *Handbook of Self-regulation: Research, Theory, and Applications*. New York, NY: The Guilford Press

Baumert, J., Kunter, M., Blum, W., Brunner, M., Voss, T., Jordan, A., Klusmann, U., Krauss, S., Neubrand, M., and Tsai, Y. (2010). Teachers' mathematical knowledge, cognitive activation in the classroom, and student progress. *American Educational Research Journal, 47*(1), 133-180. Available: https://doi.org/10.3102/0002831209345157 [April 2019].

Beaton, A.E., Rogers, A.M., González, E., Hanly, M.B., Kolstad, A., Rust, K.F., Sikali, E., Stokes, L., and Jia, Y. (2011). *The NAEP Primer* (NCES 2011-463). National Center for Education Statistics. Washington, DC: U.S. Department of Education.

Belfield, C.R., and Levin, H.M. (2007). *The Price We Pay: Economic and Social Consequences of Inadequate Education*. Washington, DC: Brookings. Available: https://www.brookings.edu/book/the-price-we-pay [January 2019].

Berliner, D. (2013). Effects of inequality and poverty vs. teachers and schooling on America's youth. *Teacher's College Record, 115*, 1-26.

Bierman, K.L., Nix, R.L., Greenberg, M.T., Blair, C., and Domitrovich, C.E. (2008). Executive functions and school readiness intervention: impact, moderation, and mediation in the Head Start REDI program. *Development and Psychopathology, 20*(3), 821-843.

Bill and Melinda Gates Foundation. (2010). *Learning about Teaching: Initial Findings from the Measures of Effective Teaching Project.* Seattle: Author.

Bireda, S., and Chait, R. (2011). *Increasing Teacher Diversity: Strategies to Improve the Teaching Workforce.* Washington, DC: Center for American Progress. Available: https://files.eric.ed.gov/fulltext/ED535654.pdf [August 2019].

Black, D. (2017). *Education Law: Equality, Fairness, and Reform.* New York, NY: Aspen Publishers, Inc.

Blair, C., and Diamond, A. (2008). Biological processes in prevention and intervention: The promotion of self-regulation as a means of preventing school failure. *Development and Psychopathology, 20*(3), 899-911. Available: https://doi.org/10.1017/S0954579408000436 [January 2019].

Blair, C., and Raver, C. (2012). Child development and the context of adversity: Experimental canalization of brain and behavior. *American Psychologist, 67*, 309-318.

Blair, C., and Razza, R.P. (2007). Relating effortful control, executive function, and false belief understanding to emerging math and literacy ability in kindergarten. *Child Development, 78*, 647-663. Available: https://doi.org/10.1111/j.1467-8624.2007.01019.x [January 2019].

Blazar, D., and Kraft, M. (2017). Teacher and teaching effects on students' behavior and attitudes. *Educational Evaluation and Policy Analysis, 39*(1), 146-170.

Bohrnstedt, G., Kitmitto, S., Ogut, B., Sherman, D., and Chan, D. (2015). School Composition and the Black–White Achievement Gap: Methodology Companion (NCES 2015-032). National Center for Education Statistics, Institute of Education Sciences, U.S. Department of Education, Washington, DC.

Bourque, M.L. (2009). A History of NAEP Achievement Levels: Issues, Implementation, and Impact 1989-2009. Paper commissioned for the 20th anniversary of the National Assessment Governing Board. Available: https://www.nagb.org/publications/reports-papers/achievement-levels/history-naep-achievement-levels-1989-2009.html [August 2019].

Bowen, W.G., Chingos, M.M., and McPherson, M.S. (2009). *Crossing the Finish Line: Completing College at America's Public Universities.* Princeton, NJ: Princeton University Press.

Bowers, A.J. (2010). Analyzing the longitudinal K–12 grading histories of entire cohorts of students: Grades, data driven decision making, dropping out and hierarchical cluster analysis. *Practical Assessment, Research, & Evaluation, 15*(7), 1-8. Available: https://pareonline.net/pdf/v15n7.pdf [April 2019].

Bowers, H., Manion, I., Papadopoulos, D., and Gauvreau, E. (2013). Stigma in school-based mental health: Perceptions of young people and service providers. *Child and Adolescent Mental Health, 18*(3), 165-170.

Boyd, D., Grossman, P., Lankford, H., Loeb, S., and Wyckoff, J. (2008). *Measuring Effect Sizes: The Effect of Measurement Error.* Working Paper 19. Washington, DC: National Center for Analysis of Longitudinal Data in Education Research, The Urban Institute.

Braxton, J.M. (2000). Reinvigorating theory and research on the departure puzzle. In J.M. Braxton (Ed.), *Reworking the Student Departure Puzzle* (pp. 257-274). Nashville, TN: Vanderbilt University Press.

Brock, S.E., Nickerson, A.B., Reeves, M.A., Jimerson, S.R., Feinberg, T., and Lieberman, R. (2009). School crisis prevention and intervention: The PREPaRE model. Bethesda, MD: National Association of School Psychologists.

Brookover, W.B., Beady, C., Flood, P., Schweitzer, J., and Wisenbaker, J. (1979). *School Social Systems and Student Achievement: Schools Can Make a Difference.* New York, NY: Praeger Publishers.

Brooks-Gunn, J., and Duncan, G.J. (1997). The effects of poverty on children. *The Future of Children, 7*(2), 55-71. Available: https://www.researchgate.net/publication/13921271_The_Effects_of_Poverty_on_Children [December 2018].

Brooks-Gunn, J., Duncan, G.J., and Aber, J.L. (Eds.). (1997). *Neighborhood Poverty: Context and Consequences for Children.* New York, NY: Russell Sage Foundation.

Brooks-Gunn, J., and Markman, L.B. (2005). The contribution of parenting to ethnic and racial gaps in school readiness. *The Future of Children, 15*(1), 139-168. Available: https://files.eric.ed.gov/fulltext/EJ795847.pdf [April 2019].

Bruch, S.K., and Soss, J. (2018). Schooling as a formative political experience: Authority relations and the education of citizens. *Perspectives on Politics, 16*(01), 36-57. Available: https://doi.org/10.1017/S1537592717002195 [March 2019].

Bryk, A.S., and Hermanson, K.L. (1993). Chapter 10: Educational indicator systems: Observations on their structure, interpretation, and use. *Review of Research in Education, 19*(1), 451-484.

Bryk, A.S., Sebring, P.B., Allensworth, E., Luppescu, S., and Easton, J.Q. (2010). *Organizing Schools for Improvement: Lessons from Chicago.* Chicago, IL: University of Chicago Press.

Bryk, A.S., and Thum, Y.M. (1989). The effects of high school organization on dropping out: An exploratory investigation. *American Educational Research Journal, 26*(3), 353-383.

Burchinal, M., Kainz, K., and Cai, Y. (2011). How well do our measures of quality predict child outcomes? A meta-analysis and coordinated analysis of data from large-scale studies of early childhood settings. In M. Zaslow, I. Martinez-Beck, K. Tout, and T. Halle (Eds.), *Quality Measurement in Early Childhood Settings* (pp. 11-31). Baltimore, MD: Paul H Brookes Publishing.

Burchinal, M., Magnuson, K., Powell, D., and Hong, S.S. (2015). Early childcare and education. In M.H. Bornstein, T. Leventhal, and R.M. Lerner (Eds.), *Handbook of Child Psychology and Developmental Science: Ecological Settings and Processes* (pp. 223-267). Hoboken, NJ: John Wiley & Sons, Inc.

Burchinal, M., Vandergrift, N., Pianta, R., and Mashburn, A. (2010). Threshold analysis of association between child care quality and child outcomes for low-income children in pre-kindergarten programs. *Early Childhood Research Quarterly, 25*(2), 166-176. Available: http://dx.doi.org/10.1016/j.ecresq.2009.10.004 [April 2019].

Buysse, V., and Peisner-Feinberg, E. (2010). Recognition & response: Response to intervention for PreK. *Young Exceptional Children, 13*(4), 2-13. Available: https://doi.org/10.1177/1096250610373586 [April 2019].

Callahan, R.M., and Shifrer, D. (2016). Equitable access for secondary English learner students: Course taking as evidence of EL program effectiveness. *Educational Administration Quarterly, 52*(3), 463-496.

Camara, W.J., and Echternacht, G. (2000). *The SAT I and High School Grades: Utility in Predicting Success in College* (Research Notes RN-10). The College Board, Office of Research and Development. Available: http://research.collegeboard.org/sites/default/files/publications/2012/7/researchnote-2000-10-sat-high-school-grades-predicting-success.pdf [January 2019].

Campbell, D.E. (2006). *Why We Vote: How Schools and Communities Shape Our Civic Life.* Princeton, NJ: Princeton University Press. Available: http://site.ebrary.com/id/10443117 [March 2019].

Campbell, D.E. (2007). Sticking together: Classroom diversity and civic education. *American Politics Research, 35*(1), 57-78.

Campbell, D.E. (2008). Voice in the classroom: How an open classroom climate fosters political engagement among adolescents. *Political Behavior, 30*(4), 437-454.

Campbell, D.E. (2018). *Measuring the Civic Participation of Adolescents.* Paper prepared for the Committee on Developing Indicators of Educational Equity. Available: http://sites.nationalacademies.org/cs/groups/dbassesite/documents/webpage/dbasse_193230.pdf [May 2019].

Cantor, P., Osher, D., Berg, J., Strayer, L., and Rose, T. (2018). Malleability, plasticity, and individuality: How children learn and develop in context. *Applied Developmental Science.* Available: https://www.tandfonline.com/doi/full/10.1080/10888691.2017.1398649 [May 2019].

Cantrell, S., and Kane, T. (2013). *Ensuring Fair and Reliable Measures of Effective Teaching: Culminating Findings from the MET Project's Three-Year Study.* Seattle: Bill and Melinda Gates Foundation. Available: http://k12education.gatesfoundation.org/resource/ensuring-fair-and-reliable-measures-of-effective-teaching-culminating-findings-from-the-met-projects-three-year-study/ [August 2019].

Capra, F. (1982). *The Turning Point: Science, Society, and the Rising Culture.* New York, NY: Simon and Schuster.

Carnevale, A.P., Smith, N., and Melton, M. (2011). *STEM: Science Technology Engineering Mathematics. State-Level Analysis.* Washington, DC: Georgetown University Center on Education and the Workforce. Available: https://eric.ed.gov/?id=ED525307 [January 2019].

Casserly, M., Price-Baugh, R., Corcoran, A., Lewis, S., Uzzell, R., Simon, C., Heppen, J., Leinwand, S., Salinger, T., Bandeira de Mello, V., Dogan, E., and Novotny, L. (2011). *Pieces of the Puzzle: Factors in the Improvement of Urban School Districts on the National Assessment of Educational Progress.* Washington, DC: Council of the Great City Schools.

Castellano, K.E., and Ho., A.D. (2013). *A Practitioner's Guide to Growth Models.* Washington, DC: Council of Chief State School Officers. Available: https://scholar.harvard.edu/files/andrewho/files/a_pracitioners_guide_to_growth_models.pdf [December 2018].

Centers for Disease Control and Prevention. (2008). *National Health Interview Survey 2007.* Atlanta: Author.

Chang, H.N., Bauer, L., and Byrnes, V. (2018). *Data Matters: Using Chronic Absence to Accelerate Action for Student Success.* Attendance Works and Everyone Graduates Center. Available: https://www.attendanceworks.org/wp-content/uploads/2018/09/Data-Matters_090618_FINAL.pdf [November 2018].

Chetty, R., Friedman, J.N., and Rockoff, J.E. (2011). *New Evidence on the Long-Term Impacts of Tax Credits.* IRS Statistics of Income White Paper. Washington, DC: Internal Review Service. Available: https://www.irs.gov/pub/irs-soi/11rpchettyfriedmanrockoff.pdf [April 2019].

Chetty, R., Friedman, J.N., and Rockoff, J.E. (2014). Measuring the impacts of teachers II: Teacher value-added and student outcomes in adulthood. *American Economic Review, 104*(9), 2633-2679.

Chetty, R., Hendren, N., and Katz, L.F. (2016). The effects of exposure to better neighborhoods on children: New evidence from the Moving to Opportunity experiment. *American Economic Review, 106*(4), 855-902.

Child Trends. (2015). *Early School Readiness: Indicators of Child and Youth Well-Being.* Bethesda, MD: Author.

Citro, C.F. (2012). Content. In M.J. Anderson, C.F. Citro, and J.J. Salvo (Eds.), *Encyclopedia of the U.S. Census—From the Constitution to the American Community Survey (ACS),* 2nd Edition (pp. 102-104). Washington, DC: CQPress.

Clements, D.H., and Sarama, J. (2008). Experimental evaluation of the effects of a research-based preschool mathematics curriculum. *American Educational Research Journal, 45*(2), 443-494. Available: https://doi.org/10.3102/0002831207312908 [April 2019].

Clewell, B.C., Puma, M.J., and McKay, S.A. (2001). *Does It Matter If My Teacher Looks Like Me?: The Impact of Teacher Race and Ethnicity on Student Academic Achievement.* Washington, DC: Urban Institute, Education Policy Center.

Clotfelter, C.T., Ladd, H.F., and Vigdor, J.L. (2010). Teacher credentials and student achievement in high school: A cross-subject analysis with student fixed effects. *Journal of Human Resources, 45*(3), 655-681.

Clotfelter, C.T., Ladd, H.F., and Vigdor, J.L. (2015). The aftermath of accelerating algebra: Evidence from district policy initiatives. *Journal of Human Resources, 50*(1), 159-188.

Clotfelter, C.T., Ladd, H.F., Vigdor, J.L., and Wheeler, J. (2007). High poverty schools and the distribution of teachers and principals. *North Carolina Law Review, 85*(5), 1345-1380.

Cohen, J., and Geier, V.K. (2010). School climate research summary- January 2010. *School Climate Brief, 1*(1), 1-16. Available: http://community-matters.org/downloads/SchoolClimateChangeJan2010.pdf [April 2019].

Coleman, J.S. (1988). Social capital in the creation of human capital. *American Journal of Sociology, 94*(Suppl.), S95-S120.

Coleman, J.S. (1990). *Foundations of Social Theory.* Cambridge, MA: Belknap Press of Harvard University Press.

Coleman, J.S., and Hoffer, T. (1987). *Public and Private High Schools: The Impact of Communities.* New York, NY: Basic Books.

Condron, D., Tope, D., Steidl, C.R, and Freeman, K.J. (2013). Racial segregation and the black/white achievement gap 1992-2009. *The Sociological Quarterly 54*(1), 130-157.

Conger, D. (2005). Within-school segregation in an urban school district. *Educational Evaluation and Policy Analysis, 27*(3), 225-244. Available: https://doi.org/10.3102/01623737027003225 [April 2019].

Conley, D.T. (2007). *Redefining College Readiness.* Eugene, OR: Educational Policy Improvement Center.

Conley, D.T. (2008). Rethinking college readiness. *New Directions in Higher Education, 2008*(144), 3-13.

Conley, D.T., and French, E.M. (2014). Ownership of learning and self-efficacy as key components of college readiness. *American Behavioral Scientist, 58*(8), 1018-1034.

Conley, D.T., Beach, P., Thier, M., Lench, S.C., and Chadwick, K.L. (2014). *Measures for a College and Career Indicator: Final Report.* Eugene, OR: Educational Policy Improvement Center.

Connell, J.P., and Wellborn, J.G. (1991). Competence, autonomy, and relatedness: A motivational analysis of self-system processes. In M.R. Gunnar and L.A. Sroufe (Eds.), *Self-Processes and Development: Minnesota Symposium on Child Psychology* (Vol. 23, pp. 43-77). Chicago, IL: University of Chicago Press.

Corcoran, S.P., and Evans, W.N. (2008). Stalled progress in closing the race achievement gap: The role of teacher quality. In K. Magnuson and J. Waldfogel, *Steady Gains and Stalled Progress: Inequality and the Black-White Test Score Gap.* New York, NY: Russell Sage Foundation Press.

Corno, L., and Mandinach, E.B. (1983). The role of cognitive engagement in classroom learning and motivation. *Educational Psychologist, 18*(2), 88-108.

Cortes, K.E., Goodman, J.S., and Nomi, T. (2015). Intensive math instruction and educational attainment: Long-run impacts of double-dose algebra. *Journal of Human Resources, 50*(1), 108-158.

Croninger, R.G., and Lee, V.E. (2001). Social capital and dropping out of high school: Benefits to at-risk students of teachers' support and guidance. *Teachers College Record, 103*(4), 548-581.

Crosnoe, R., Mistry, R.S., and Elder Jr., G.H. (2002). Economic disadvantage, family dynamics, and adolescent enrollment in higher education. *Journal of Marriage and Family, 64(3)*, 690-672.

Currie, J. (2009). Healthy, wealthy, and wise: Socioeconomic status, poor health in childhood, and human capital development. *Journal of Economic Literature, 47*(1), 87-122.

Cutler, D.M., and Lleras-Muney, A. (2006). *Education and Health: Evaluating Theories and Evidence.* NBER Working Papers 12352. Cambridge, MA: National Bureau of Economic Research.

Darling-Hammond, L., and Post, L. (2000). Inequality in teaching and schooling: Supporting high quality teaching and leadership in low income schools. In R.D. Kahlenberg (Ed.), *A Nation at Risk: Preserving Public Education as an Engine for Social Mobility.* New York, NY: The Century Foundation Press.

Darling-Hammond, L., and Wise, A.E. (1985). Beyond standardization: State standards and school improvement. *The Elementary School Journal, 85*(3), 315-336.

Data Quality Campaign. (2019). *Growth Data: It Matters and It's Complicated.* Washington, DC: Author. Available: https://dataqualitycampaign.org/resource/growth-data-it-matters-and-its-complicated [April 2019].

Dawes, N.P., and Larson, R. (2011). How youth get engaged: Grounded-theory research on motivational development in organized youth programs. *Developmental Psychology, 47*, 259-269. Available: http://dx.doi.org/10.1037/a0020729 [April 2019].

Day, J.C., and Newburger, E.C. (2002). *The Big Payoff: Educational Attainment and Synthetic Estimates of Work-Life Earnings.* Washington, DC: U.S. Census Bureau. Available: https://www.census.gov/prod/2002pubs/p23-210.pdf [April 2019].

de Brey, C., Musu, L., McFarland, J., Wilkinson-Flicker, S., Diliberti, M., Zhang, A., Branstetter, C., and Wang, X. (2019). *Status and Trends in the Education of Racial and Ethnic Groups 2018* (NCES 2019-038). Washington, DC: U.S. Department of Education, National Center for Education Statistics. Available: https://nces.ed.gov/pubs2019/2019038.pdf [April 2019].

DeAngelis, K.J., and Presley, J.B. (2011). Teacher qualifications and school climate: Examining their interrelationship for school improvement. *Leadership and Policy in Schools, 10*(1), 84-120.

Dearing, E., and Taylor, B.A. (2007). Home improvements: Within-family associations between income and the quality of children's home environments. *Journal of Applied Developmental Psychology, 28*, 427-444.

Dee, T.S. (2004). Teachers, race, and student achievement in a randomized experiment. *Review of Economics and Statistics, 86*(1), 195-210.

Dee, T.S., Jacob, B., and Schwartz, N.L. (2013). The effects of NCLB on school resources and practices. *Educational Evaluation and Policy Analysis, 35*(2), 252-279. Available: https://doi.org/10.3102/0162373712467080 [April 2019].

Deming, D.J. (2015). *The Growing Importance of Social Skills in the Labor Market.* Working Paper No. 21472. Cambridge, MA: National Bureau of Economic Research.

Dilworth-Bart, J.E., and Moore, C.F. (2006). Mercy mercy me: Social injustice and the prevention of environmental pollutant exposures among ethnic minority and poor children. *Child Development, 77*(2), 247-265.

DiPrete, T.A., and Buchmann, C. (2013). *The Growing Gender Gap in Education and What It Means for American Schools.* New York, NY: Russell Sage Foundation.

Domina, T., McEachin, A., Penner, A., and Penner, E. (2015). Aiming high and falling short: California's eighth-grade algebra-for-all effort. *Educational Evaluation and Policy Analysis, 37*(3), 275-295.

Dougherty, S.M., Goodman, J.S., Hill, D.V., Litke, E.G., and Page, L.C. (2015). Middle school math acceleration and equitable access to eighth-grade algebra: Evidence from the Wake County Public School System. *Educational Evaluation and Policy Analysis, 37*(1 Suppl.), 80S-101S.

Dougherty, C., Mellor, L., and Jian, S. (2006). *The Relationship between Advanced Placement and College Graduation.* 2005 AP Study Series, Report 1. Austin, TX: National Center for Educational Accountability. Available: https://eric.ed.gov/?id=ED519365 [April 2019].

Duckworth, A.L., and Seligman, M.E.P. (2005). Self-discipline outdoes IQ in predicting academic performance of adolescents. *Psychological Science, 16,* 939-944. Available: doi:10.1111/j.1467-9280.2005.01641.x [April 2019].

Duckworth, A.L., Peterson, C., Matthews, M.D., and Kelly, D.R. (2007). Grit: Perseverance and passion for long-term goals. *Journal of Personality and Social Psychology, 92*(6), 1087-1101.

Duncan, G.J., and Brooks-Gunn, J. (Eds.) (1999). *Consequences of Growing Up Poor.* New York, NY: Russell Sage Foundation.

Duncan, G.J., and Magnuson, K. (2011). The nature and impact of early achievement skills, attention and behavior problems. In G.J. Duncan and R.J. Murnane (Eds.), *Whither Opportunity: Rising Inequality, Schools, and Children's Life Chances* (pp. 47-69). New York, NY: Russell Sage Foundation.

Duncan, G.J., and Murnane, R. (Eds.). (2011). *Whither Opportunity? Rising Inequality, Schools, and Children's Life Chances.* New York, NY: Russell Sage Foundation.

Duncan, G.J., Brooks Gunn, J., and Klebanov, P.K. (1994). Economic deprivation and early childhood development. *Child Development, 65*(2), 296-318.

Duncan, G.J., Dowsett, C.J., Claessens, A., Magnuson, K., Huston, A.C., Klebanov, P., Pagani, L.S., Feinstein, L., Engel, M., Brooks-Gunn, J., Sexton, H., Duckworth, K., and Japel, C. (2007). School readiness and later achievement. *Developmental Psychology, 43*(6), 1428-1446. Available: https://www.ncbi.nlm.nih.gov/pubmed/18020822 [August 2017].

Duncombe, W., and Yinger, J. (2005). How much more does a disadvantaged student cost? *Economics of Education Review, 24*(5), 513-532.

Easton, J.Q., Johnson, E., and Sartain, L. (2017). *The Predictive Power of Ninth-Grade GPA.* Chicago, IL: University of Chicago Consortium on School Research. Available: https://consortium.uchicago.edu/sites/default/files/2018-10/Predictive%20Power%20of%20Ninth-Grade-Sept%202017-Consortium.pdf [April 2019].

Educational Testing Service. (2013). *Poverty and Childhood Education: Finding a Way Forward.* Princeton, NJ: Author.

Egalite, A.J., and Kisida, B. (2017). The effects of teacher match on students' academic perceptions and attitudes. *Educational Evaluation and Policy Analysis, 40*(1), 59-81.

Egalite, A.J., Kisida, B., and Winters, M.A. (2015). Representation in the classroom: The effect of own-race teachers on student achievement. *Economics of Education Review, 45,* 44-52.

Ehrlich, S.B., Gwynne, J.A., and Allensworth, E.M. (2018). Pre-kindergarten attendance matters: Early chronic absence patterns and relationships to learning outcomes. *Early Childhood Research Quarterly, 44,* 136-151.

Elliott, D.S., Wilson, W.J., Huizinga, D., Sampson, R.J., Elliott, A., and Rankin, B. (1996). The effects of neighborhood disadvantage on adolescent development. *Journal of Research in Crime and Delinquency, 33*(4), 389-426. Available: https://doi.org/10.1177/0022427896033004002 [April 2019].

Evans, G.W. (2004). The environment of childhood poverty. *American Psychologist, 59*, 77-92.

ExcelinEd. (2018). *College and Career Pathways: Equity and Access.* Tallahassee, FL: Author. Available: https://www.excelined.org/wp-content/uploads/2018/10/ExcelinEd.Report. CollegeCareerPathways.CRDCAnalysis.2018.pdf [April 2019].

Fantuzzo, J.W., Gadsden, V.L., and McDermott, P.A. (2010). An integrated curriculum to improve mathematics, language, and literacy for Head Start children. *American Educational Research Journal, 48*(3), 763-793.

Farkas, G. (2003). Racial disparities and discrimination in education: What do we know, how do we know it, and what do we need to know? *Teachers College Record, 105*(6), 1119-1146.

Farrington, C.A. (2013). *Academic Mindsets as a Critical Component of Deeper Learning.* Menlo Park, CA: William and Flora Hewlett Foundation. Available: https://www.hewlett. org/wp-content/uploads/2016/08/Academic_Mindsets_as_a_Critical_Component_of_ Deeper_Learning_CAMILLE_FARRINGTON_April_20_2013.pdf [April 2019].

Farrington, C.A. (2017). *Becoming Effective Learners (BEL-S): Looking at Within-Student Differences in Self-Report across Classroom Contexts.* Presentation prepared for the Conference on Measuring and Assessing Skills 2017, Chicago, IL. Research Network on the Determinants of Life Course Capabilities and Outcomes.

Farrington, C.A., Roderick, M., Allensworth, E., Nagaoka, J., Keyes, T.S., Johnson, D., and Beechum, N.O. (2012). *Teaching Adolescents to Become Learners: The Role of Noncognitive Factors in Shaping School Performance.* Chicago, IL: University of Chicago Consortium on Chicago School Research.

Faster, D., and Lopez, D. (2013). School climate and assessment. In T. Dary and T. Pickeral, (Eds.), *School Climate Practices for Implementation and Sustainability. A School Climate Practice Brief, Number 1.* New York, NY: National School Climate Center. Available: http://www.schoolclimate.org/publications/practice-briefs.php [April 2019].

Federal Interagency Forum on Child and Family Statistics. (2012a). *ED1 Family Reading to Young Children: Percentage of Children Ages 3–5 Who Were Read to Every Day in the Last Week by a Family Member by Child and Family Characteristics, and Region, Selected Years 1993–2007.* Available: http://childstats.gov/americaschildren/tables/ed1.asp? popup=true [April 2019].

Federal Interagency Forum on Child and Family Statistics. (2012b). FAM1.B *Family Structure and Children's Living Arrangements: Detailed Living Arrangements of Children by Gender, Race and Hispanic Origin, Age, Parent's Education, and Poverty Status, 2011.* Available: http://childstats.gov/americaschildren/tables/fam1b.asp?popup=true [April 2019].

Federal Interagency Forum on Child and Family Statistics. (2016). *America's Children in Brief: Key National Indicators of Well-Being, 2016.* Washington, DC: U.S. Government Printing Office.

Federal Interagency Forum on Child and Family Statistics. (2017). *America's Children: Key National Indicators of Well-Being, 2017.* Washington, DC: U.S. Government Printing Office.

Ferguson, T.J., Stegge, H. and Damhuis, I. (1991). Children's understanding of guilt and shame. *Child Development, 62*, 827-839. Available: doi:10.1111/j.1467-8624.1991. tb01572.x [April 2019].

Field, S., Kuczera, M., and Pont, B. (1988). *No More Failures: Ten Steps to Equity in Education.* Paris: OECD. Available: https://www.oecd.org/education/school/45179151.pdf [August 2019].

Finn, J.D. (1989). Withdrawing from school. *Review of Educational Research, 59*(2), 117-142.

Finn, J.D., Pannozzo, G.M., and Voelkl, K.E. (1995). Disruptive and inattentive-withdrawn behavior and achievement among fourth graders. *The Elementary School Journal, 95*(5), 421-454.

Finn, J.D., and Rock, D.A. (1997). Academic success among students at risk for school failure. *Journal of Applied Psychology, 82*(2), 221-234.

Firestone, W.A., Mayrowetz, D., and Fairman, J. (1998). Performance-based assessment and instructional change: The effects of testing in Maine and Maryland. *Educational Evaluation and Policy Analysis, 20*(2), 95-113.

Flannery, M.E. (2015). The school-to-prison pipeline: Time to shut it down. *neaToday*, January 5. Available: http://neatoday.org/2015/01/05/school-prison-pipeline-time-shut [August 2019].

Fouts, J.T., and Myers, R.E. (1992). Classroom environments and middle school students' views of science. *The Journal of Educational Research, 85*(6), 356-361. Available: http://dx.doi.org/10.1080/00220671.1992.9941138 [April 2019].

Fox, L., Carta, J.J., Strain, P.S., Dunlap, G., and Hemmeter, M.L. (2010). Response to intervention and the pyramid model. *Infants and Young Children, 23*(1), 3-13.

Fraga, L.R., Meier, K.J., and England, R.E. (1986). Hispanic Americans and educational policy: Limits to equal access. *The Journal of Politics, 48*(4), 850-876.

Frederiksen, J.R., and Collins, A. (1989). A systems approach to educational testing. *Educational Researcher, 18*(9), 27-32.

Fredricks, J.A., Blumenfeld, P.C., and Paris, A. (2004). School engagement: Potential of the concept: State of the evidence. *Review of Educational Research, 74*, 59-119. Available: doi: 10.3102/00346543074001059 [April 2019].

Fredricks, J.A., and McColskey, W. (2012). The measurement of student engagement: A comparative analysis of various methods and student self-report instruments. In S.L. Christenson et al. (Eds.), *Handbook of Research on Student Engagement* (pp. 763-782). Available: https://www.lcsc.org/cms/lib/MN01001004/Centricity/Domain/108/The%20Measurement%20of%20Student%20Engagement-%20A%20Comparative%20Analysis%20of%20Various%20Methods.pdf [April 2019].

Friedman-Krauss, A.H., Barnett, W.S., Weisenfeld, G.G., Kasmin, R., DiCrecchio, N., and Horowitz, M. (2018). *The State of Preschool 2017: State Preschool Yearbook*. New Brunswick, NJ: National Institute for Early Education Research. Available: http://nieer.org/wp-content/uploads/2019/02/State-of-Preschool-2017-Full-2-13-19_reduced.pdf [April 2019].

Fronius, T., Persson, H., Guckenburg, S., Hurley, N., and Petrosino, A. (2016). *Restorative Justice in U.S. Schools: A Research Review*. San Francisco, CA: WestEd.

Fuhs, M.W., Nesbitt, K.T., Farran, D.C., and Dong, N. (2014). Longitudinal associations between executive functioning and academic skills across content areas. *Developmental Psychology, 50*, 1698-1709. Available: http://dx.doi.org/10.1037/a0036633 [April 2019].

Gamoran, A. (1987). The stratification of high school learning opportunities. *Sociology of Education, 60*(3), 135-155.

Gamoran, A., and Mare, R.D. (1989). Secondary school tracking and educational inequality: Compensation, reinforcement, or neutrality? *American Journal of Sociology, 94*(5), 1146-1183.

Gamoran, A., Porter, A.C., Smithson, J., and White, P.A. (1997). Upgrading high school mathematics instruction: Improving learning opportunities for low-achieving, low-income youth. *Educational Evaluation and Policy Analysis, 19*(4), 325-338. Available: https://journals.sagepub.com/doi/10.3102/01623737019004325 [April 2019].

Gandara, P., and Rumberger, R.W. (2008). Defining an adequate education for English learners. *Education Finance and Policy, 3*(1), 130-148.

Gao, N. (2016). *College Readiness in California: A Look at Rigorous High School Course-Taking*. San Francisco, CA: Public Policy Institute of California. Available: https://www.ppic.org/content/pubs/report/R_0716NGR.pdf [December 2018].

Garrett, P., Ng'andu, N., and Ferron, J. (1994). Poverty experience of young children and the quality of their home environments. *Child Development, 65*, 331-345.

Gee, J.P. (2017). Identity and diversity in today's world. *Multicultural Education Review*, 9(2), 83-92. Available: https://www.tandfonline.com/doi/full/10.1080/2005615X.2017.1312216 [April 2019].

Geiser, S., and Santelices, M.V. (2007). *Validity of High-School Grades in Predicting Student Success beyond the Freshman Year: High-School Record vs. Standardized Tests as Indicators of Four-Year College Outcomes*. Research & Occasional Paper Series: CSHE.6.07. Berkeley, CA: Center for Studies in Higher Education, University of California, Berkeley. Available: https://cshe.berkeley.edu/sites/default/files/publications/rops.geiser._sat_6.13.07.pdf [April 2019].

Geiser, S., and Studley, R. (2002). UC and the SAT: Predictive validity and differential impact of the SAT I and SAT II at the University of California. *Educational Assessment*, 8(1), 1-26. Available: http://dx.doi.org/10.1207/S15326977EA0801_01 [April 2019].

George, J., and Darling-Hammond, L. (2019). *The Federal Role and School Integration: Brown's Promise and Present Challenges*. Palo Alto, CA: Learning Policy Institute.

Gershenson, S., Jacknowitz, A., and Brannegan, A. (2017). Are student absences worth the worry in U.S. primary schools? *Education Finance and Policy*, 12(2), 137-165.

Gershenson, S., Hart, C.M.D., Lindsay, C.A., Papageorge, N.W. (2018). *The Long-Run Impacts of Same-Race Teachers*. NBER Working Paper No. 25254. Cambridge, MA: National Bureau of Economic Research. Available: https://www.nber.org/papers/w25254 [April 2019].

Gershenson, S., Holt, S.B., and Papageorge, N.W. (2016). Who believes in me? The effect of student-teacher demographic match on teacher expectations. *Economics of Education Review*, 52, 209-224.

Gimpel, J.G., Schuknecht, J.E., and Lay, J.C. (2003). *Cultivating Democracy: Civic Environments and Political Socialization in America*. Washington, DC: Brookings Institution Press.

Ginsburg, A.L., Noell, J., and Plisko, V.W. (1988). Lessons from the wall chart. *Educational Evaluation and Policy Analysis*, 10(1), 1-12. doi: 10.2307/1163860.

Gitomer, D. and Zisk, R. (2015, March). Knowing what students know. *Review of Research in Education*, 39(n), 1-53. doi: 10.3102/0091732X14557001.

Glancy, E., Fulton, M., Anderson, A., Zinth, J., Millard, M., and Delander, B. (2014). *Blueprint for College Readiness*. Denver, CO: Education Commission of the States.

Glaser, R., Linn, R., and Bohrnstedt, G. (1997). *Assessments in Transition: Monitoring the Nation's Educational Progress*. Stanford, CA: National Academy of Education.

Goldhaber, D.D., and Brewer, D.J. (2000). Does teacher certification matter? High school teacher certification status and student achievement. *Educational Evaluation and Policy Analysis*, 22(2), 129-145.

Goldhaber, D., and Hansen, M. (2010). Using performance on the job to inform teacher tenure decisions. *American Economic Review*, 100(2), 250-255.

Goldhaber, D., Lavery, L., and Theobald, R. (2015). Uneven playing field? Assessing the teacher quality gap between advantaged and disadvantaged students. *Educational Researcher*, 44(5), 293-307.

Goldin, C., and Katz, L.F. (2010). *The Race between Education and Technology*. Cambridge, MA: Harvard University Press.

González, T. (2012). Keeping kids in schools: Restorative justice, punitive discipline, and the school to prison pipeline. *Journal of Law & Education*, 41(2), 281-335. Available at SSRN: https://ssrn.com/abstract=2658513 [August 2019].

Goodman, J. (2014). *Flaking Out: Student Absences and Snow Days as Disruptions of Instructional Time*. NBER Working Paper No. w20221. Cambridge, MA: National Bureau of Economic Research.

Gormley, W.T., Jr., Gayer, T., Phillips, D., and Dawson, B. (2005). The effects of universal pre-K on cognitive development. _Developmental Psychology, 41_(6), 872-884.

Gottfried, M.A. (2009). Excused versus unexcused: How student absences in elementary school affect academic achievement. _Educational Evaluation and Policy Analysis, 31_(4), 392-415.

Gottfried, M.A. (2010). Evaluating the relationship between student attendance and achievement in urban elementary and middle schools: An instrumental variables approach. _American Educational Research Journal, 47_(2), 434-465. Available: https://www.jstor.org/stable/40645446?seq=1#page_scan_tab_contents [April 2019].

Gottfried, M.A. (2014). Chronic absenteeism and its effects on students' academic and socioemotional outcomes. _Journal of Education for Students Placed at Risk, 19_(2), 53-75. Available: https://doi.org/10.1080/10824669.2014.962696 [April 2019].

Gottfried, M.A. (2017). Linking getting to school with going to school. _Educational Evaluation and Policy Analysis, 39_(4), 571-592.

Gottfried, M.A. (2019). Chronic absenteeism in the classroom context: Effects on achievement. _Urban Education, 54_(1), 3-34. Available: https://doi.org/10.1177/0042085915618709 [April 2019].

Grant, S., Hamilton, L.S., Wrabel, S.L., Whitaker, A., Gomez, C., Leschitz, J.T., Unlu, F., Chavez-Herrerias, E., Baker, G., Barrett, M., Harris, M., and Ramos, A. (2017). _Social and Emotional Learning Interventions Under the Every Student Succeeds Act: Evidence Review_ (RR-2133-WF). Santa Monica, CA: RAND.

Grissom, J.A., and Redding, C. (2016). Discretion and disproportionality. _AERA Open, 2_(1), 233285841562217.

Gronberg, T.J., Jansen, D.W., and Taylor, L.L. (2011). The adequacy of educational cost functions: Lessons from Texas. _Peabody Journal of Education, 86_(1), 3-27.

Guryan, J., Hurst, E., and Kearney, M. (2008). Parental education and parental time with children. _Journal of Economic Perspectives, 22_(3), 23-46.

Gwynne, J., Lesnick, J., Hart, H.M., and Allensworth, E.M. (2009). _What Matters for Staying On-Track and Graduating in Chicago Public Schools: A Focus on Students with Disabilities_. Research Report. UChicago Consortium on Chicago School Research. Available: https://consortium.uchicago.edu/publications/what-matters-staying-track-and-graduating-chicago-public-schools-focus-students [May 2019].

Hackman, D.A., Farah, M.J., and Meaney, M.J. (2010). Socioeconomic status and the brain: Mechanistic insights from human and animal research. Nature reviews. _Neuroscience, 11_(9), 651-659. Available: https://www.ncbi.nlm.nih.gov/pmc/articles/PMC2950073 [April 2019].

Halle, T., Forry, N., Hair, E.C., Perper, K., Wandner, L.D., and Whittaker, J.V. (2009). _Disparities in Early Learning and Development: Lessons from the Early Childhood Longitudinal Study–Birth Cohort (ECLS-B): Executive Summary_. Bethesda, MD: Child Trends. Available: https://www.childtrends.org/wp-content/uploads/2013/05/2009-52DisparitiesELExecSumm.pdf [April 2019].

Hanushek, E.A., Kain, J.F., and Rivkin, S.G. (2004). Why public schools lose teachers. _Journal of Human Resources, 39_(2), 326-354.

Hanushek, E.A., and Rivkin, S.G. (2006). School Quality and the Black–White Achievement Gap. NBER Working Paper 12651. National Bureau of Economic Research, Cambridge, MA. Available: http://www.nber.org/papers/w12651.pdf [August 2019].

Hanushek, E.A., and Rivkin, S.G. (2010). Generalizations about using value-added measures of teacher quality. _American Economic Review, 100_(2), 267-271.

Hanushek, E.A., and Rivkin, S.G. (2012). The distribution of teacher quality and implications for policy. _Annual Review of Economics, 4_(1), 131-157.

Harlow, C.W. (2003). *Education and Correctional Populations. Bureau of Justice Statistics Special Report.* Washington, DC: U.S. Department of Justice. Available: https://files.eric. ed.gov/fulltext/ED477377.pdf [April 2019].

Hartman, J., Wilkins, C., Gregory, L., Gould, L.F., and D'Souza, S. (2011). *Applying an On-Track Indicator for High School Graduation: Adapting the Consortium on Chicago School Research Indicator for Five Texas Districts.* Issues & Answers Report, REL 2011–No. 100. Washington, DC: U.S. Department of Education, Institute of Education Sciences, National Center for Education Evaluation and Regional Assistance, Regional Educational Laboratory Southwest. Available: http://ies.ed.gov/ncee/edlabs [April 2019].

Haynes, N.M., Emmons, C., and Ben-Avie, M. (1997). School climate as a factor in student adjustment and achievement. *Journal of Educational and Psychological Consultation, 8*(3), 321-329.

Heckman, J.J., and LaFontaine, P.A. (2007). The American High School Graduation Rate: Trends and Levels. NBER Working Paper 13670, National Bureau of Economic Research, Cambridge, MA. Available: http://ftp.iza.org/dp3216.pdf [August 2019].

Heckman, J.J., and LaFontaine, P.A. (2010). The American high school graduation rate: Trends and levels. *The Review of Economics and Statistics, 92*(2), 244-262.

Heppen, J.B., Walters, K., Clements, M., Faria, A.-M., Tobey, C., Sorensen, N., and Culp, K. (2011). *Access to Algebra I: The Effects of Online Mathematics for Grade 8 Students.* Washington, DC: U.S. Department of Education. Available: papers3://publication/uuid/48CED158-1852-495F-BDE9-EC86D889053A [April 2019].

Hess, D.E. (2009). *Controversy in the Classroom: The Democratic Power of Discussion.* The Critical Social Thought Series. New York, NY: Routledge.

Hess, D.E., and Posselt, J. (2002). How high school students experience and learn from the discussion of controversial public issues. *Journal of Curriculum and Supervision, 17*(4), 283-314.

Hill, H.C., Rowan, B., and Ball, D.L. (2005). Effects of teachers' mathematical knowledge for teaching on student achievement. *American Educational Research Journal, 42*(2), 371-406. Available: https://doi.org/10.3102/00028312042002371 [April 2019].

Hindman, A.H., Connor, C.M., Jewkes, A.M., and Morrison, F.J. (2008). Untangling the effects of shared book reading: Multiple factors and their associations with preschool literacy outcomes. *Early Childhood Research Quarterly, 23*(3), 330-350. Available: https://doi.org/10.1016/j.ecresq.2008.01.005 [April 2019].

Hinze-Pifer, R., and Sartain, L. (2018). Rethinking universal suspension for severe student behavior. *Peabody Journal of Education, 93*(2), 228-243.

Hiss, W.C., and Franks, V.W. (2014). *Defining Promise: Optional Standardized Testing Policies in American College and University Admissions.* Arlington, VA: The National Association for College Admission Counseling.

Ho, A.D. (2008). The problem with "proficiency": Limitations of statistics and policy under No Child Left Behind. *Educational Researcher, 37*(6), 351-360.

Holt, S.B., and Gershenson, S. (2017). The impact of demographic representation on absences and suspensions. *The Policy Studies Journal.* Available: https://doi.org/10.1111/psj.12229 [April 2019].

Hooghe, M., and Dassonneville, R. (2011). The effects of civic education on political knowledge. A two year panel survey among Belgian adolescents. *Educational Assessment, Evaluation and Accountability, 23*(4), 321-339.

Howard, S., and Sommers, S.R. (2015). Exploring the enigmatic link between religion and anti-black attitudes. *Social and Personality Psychology Compass, 9*(9), 495-510. Available: https://doi.org/10.1111/spc3.12195 [April 2019].

Hutmacher, W., Cochrane, D., and Bottani, N. (Eds.). (2001). *In Pursuit of Equity in Education.* Dordrecht: Springer Netherlands.

Huttenlocher, J., Waterfall, H., Vasilyeva, M., Vevea, J., and Hedges, L.V. (2010). Sources of variability in children's language growth. *Cognitive Psychology, 61*(4), 343-365. Available: https://www.ncbi.nlm.nih.gov/pubmed/20832781 [April 2019].

Iatarola, P., Conger, D., and Long, M.C. (2011). Determinants of high schools' advanced course offerings. *Educational Evaluation and Policy Analysis, 33*(3), 340-359. Available: https://doi.org/10.3102/0162373711398124 [April 2016].

Immordino-Yang, M.H. (2016). Emotion, sociality, and the brain's default mode network: Insights for educational practice and policy. *Policy Insights from the Behavioral and Brain Sciences, 3*(2), 211-219. Available: https://doi.org/10.1177/2372732216656869 [April 2019].

Isaacs, J.B. (2012). *Starting School at a Disadvantage: The School Readiness of Poor Children.* Washington, DC: The Brookings Institution. Available: https://www.brookings.edu/wp-content/uploads/2016/06/0319_school_disadvantage_isaacs.pdf [April 2019].

Isenberg, E., Max, J., Gleason, P., Johnson, M., Deutsch, J., Hansen, M., and Angelo, L. (2016). *Do Low-Income Students Have Equal Access to Effective Teachers? Evidence from 26 Districts.* NCEE 2017-4008. Washington, DC: U.S. Department of Education.

Jackson, C.K. (2018). *Does School Spending Matter? The New Literature on an Old Question.* NBER Working Paper No. 25368. Cambridge, MA: National Bureau of Economic Research. Available: https://www.nber.org/papers/w25368 [April 2019].

Jackson, C.K., Johnson, R.C., and Persico, C. (2016). The effects of school spending on educational and economic outcomes: Evidence from school finance reforms. *Quarterly Journal of Economics, 131*(1), 157-218.

Jacob, B.A. (2002). Where the boys aren't: Non-cognitive skills, returns to school and the gender gap in higher education. *Economics of Education Review, 21*(6), 589-598.

Jacob, B.A., and Lovett, K. (2017). *Chronic Absenteeism: An Old Problem in Search of New Answers.* Washington, DC: The Brookings Institution. Available: https://www.brookings.edu/research/chronic-absenteeism-an-old-problem-in-search-of-new-answers [April 2019].

Jacob, R.T., Stone, S., and Roderick, M. (2004). *Ending Social Promotion in Chicago: The Response of Teachers and Students.* Chicago, IL: Consortium on Chicago School Research.

Jaschik, S., and Lederman, D. (Eds.) (2018). *2018 Survey of College and University Admissions Directors: A Study by Inside Higher Ed and Gallup.* Washington, DC: Inside Higher Ed. Available: https://www.insidehighered.com/system/files/booklets/IHE_2018_Admissions_Director_Survey.pdf [April 2019].

Jennings, J.L. (2018). *Vital Signs for the American Education System.* Paper prepared for the Committee on Developing Indicators of Educational Equity. Available: http://sites.nationalacademies.org/cs/groups/dbassesite/documents/webpage/dbasse_193233.pdf [May 2019].

Jennings, J.L., and DiPrete, T.A. (2010). Teacher effects on social and behavioral skills in early elementary school. *Sociology of Education, 83*(2), 135-159.

Jez, S.J., and Wassmer, R.W. (2015). The impact of learning time on academic achievement. *Education and Urban Society, 47*(3), 284-306. Available: https://doi.org/10.1177/0013124513495275 [April 2019].

Jia, Y.N., Konold, T.R., and Cornell, D. (2016). Authoritative school climate and high school dropout rates. *School Psychology Quarterly, 31*(2), 289-303.

Jones, L.V., and Olkin, I. (Eds.) (2004). *The Nation's Report Card: Evolution and Perspectives.* Arlington, VA: Phi Delta Kappa Educational Foundation.

Jones, S.M., and Kahn, J. (2017). *The Evidence Base for How We Learn: Supporting Students' Social, Emotional, and Academic Development.* Washington, DC: The Aspen Institute. Available: https://assets.aspeninstitute.org/content/uploads/2018/03/FINAL_CDS-Evidence-Base.pdf?_ga=2.177748645.1276191698.1544659585-1494538186.1544659585 [December 2018].

Jungmann, S.M., Vollmer, N., Selby, E.A., and Witthöft, M. (2016). Understanding dys-regulated behaviors and compulsions: An extension of the emotional cascade model and the mediating role of intrusive thoughts. *Frontiers in Psychology, 7,* Article ID 994.

Kahne, J.E., and Sporte, S.E. (2008). Developing citizens: The impact of civic learning oppor-tunities on students' commitment to civic participation. *American Educational Research Journal, 45*(3), 738-766. Available: https://doi.org/10.3102/0002831208316951 [April 2019].

Kalil, A. (2015). Inequality begins at home: The role of parenting in the diverging destinies of rich and poor children. In P.R. Amato, A. Booth, S.M. McHale, and J. Van Hook (Eds.), *Families in an Era of Increasing Inequality* (pp. 63-82). New York, NY: Springer International Publishing.

Kalil, A., Ryan, R., and Corey, M. (2012). Diverging destinies: Maternal education and the developmental gradient in time with children. *Demography, 49*(4), 1361-1383.

Kane, T., Kerr, K., and Pianta, R. (2014). *Designing Teacher Evaluation Systems: New Guid-ance from the Measures of Effective Teaching Project, 1st Edition.* San Francisco, CA: Josey Bass.

Kane, T.J., Rockoff, J.E., and Staiger, D.O. (2008). What does certification tell us about teacher effectiveness? Evidence from New York City. *Economics of Education Review, 27*(6), 615-631.

Kane, T.J., and Staiger, D.O. (2008). *Estimating Teacher Impacts on Student Achievement: An Experimental Evaluation.* NBER Working Paper No. 14607. Cambridge, MA: National Bureau of Economic Research.

Kanno, Y., and Cromley, J.G. (2013). English language learners' access to and attainment in postsecondary education. *Tesol Quarterly, 47*(1), 89-121.

Kantral, T., and Kane, T. (2013). *Ensuring Fair and Reliable Measures of Effective Teaching.* Seattle: Bill and Melinda Gates Foundation.

Kaplan, S., and Kaplan, R. (1982). *Cognition and Environment: Functioning in an Uncertain World.* New York, NY: Praeger.

Karp, K. (Ed.). (2014). *Annual Perspectives in Mathematics Education 2014: Using Research to Improve Instruction.* Reston, VA: National Council of Teachers of Mathematics. Available: https://www.nctm.org/Store/Products/Annual-Perspectives-in-Mathematics-Education-2014--Using-Research-to-Improve-Instruction [April 2019].

Karp, M.M. (2011). *Toward a New Understanding of Non-Academic Student Support: Four Mechanisms Encouraging Positive Student Outcomes in the Community College.* CCRC Working Paper No. 28. New York, NY: Community College Research Center, Teachers College, Columbia University. Available: https://ccrc.tc.columbia.edu/media/k2/attachments/new-understanding-non-academic-support.pdf [April 2019].

Kelley-Kemple, T., Proger, A., and Roderick, M. (2011). *Engaging High School Students in Advanced Math and Science Courses for Success in College: Is Advanced Placement the Answer?* Evanston, IL: Society for Research on Educational Effectiveness. Available: https://eric.ed.gov/?id=ED528804 [April 2019].

Kena, G., Aud, S., Johnson, F., Wang, X., Zhang, J., Rathbun, A., Wilkinson-Flicker, S., and Kristapovich, P. (2014). *The Condition of Education 2014* (NCES 2014-083). Washington, DC: U.S. Department of Education, National Center for Education Statis-tics. Available: https://nces.ed.gov/pubs2014/2014083.pdf [April 2019].

Kena, G., Hussar W., McFarland J., de Brey C., Musu-Gillette, L., Wang, X., Zhang, J., Rathbun, A., WilkinsonFlicker, S., Diliberti M., Barmer, A., Bullock Mann, F., and Dunlop Velez, E. (2016). *The Condition of Education 2016* (NCES 2016-144). Washington, DC: U.S. Department of Education, National Center for Education Statistics. Available: https://nces.ed.gov/pubs2016/2016144.pdf [April 2019].

Kidron, Y., and Lindsay, J. (2014). *The Effects of Increased Learning Time on Student Academic and Nonacademic Outcomes: Findings from a Meta-Analytic Review* (REL 2014–015). Washington, DC: U.S. Department of Education, Institute of Education Sciences, National Center for Education Evaluation and Regional Assistance, Regional Educational Laboratory Appalachia.

Kieffer, M.J., and Marinell, W.H. (2012). *Navigating the Middle Grades: Evidence from New York City.* New York, NY: The Research Alliance for New York City Schools. Available: http://steinhardtapps.es.its.nyu.edu/ranycs/media/NavigatingMiddleGrades.pdf [April 2019].

King, K., Shumow, L., and Lietz, S. (2001). Science education in an urban elementary school: Case studies of teacher beliefs and classroom practices. *Science Education, 85*(2), 89-110.

Kinney, D.W., and Forsythe, J.L. (2005). The effects of Arts IMPACT school curriculum upon the Ohio Fourth-Grade Proficiency Test scores. *Bulletin of the Council for Research in Music Education, 164,* 35-48.

Kirsch, I., Braun, H., Lennon, M.L., and Sands, A. (2016). *Choosing our Future: A Story of Opportunity in America.* Princeton, NJ: Educational Testing Service Project.

Klugman, J., Gordon, M.F., Sebring, P.B., and Sporte, S.E. (2015). *A First Look at the 5 Essentials in Illinois Schools.* Chicago, IL: University of Chicago Consortium on Chicago School Research.

Knowles, R.T., and McCafferty-Wright, J. (2015). Connecting an open classroom climate to social movement citizenship: A study of 8th graders in Europe using IEA ICCS data. *The Journal of Social Studies Research, 39*(4), 255-269.

Koball, H., and Jiang, Y. (2018). *Basic Facts about Low-Income Children: Children under 18 Years, 2016.* New York, NY: National Center for Children in Poverty, Columbia University Mailman School of Public Health. Available: http://www.nccp.org/publications/pub_1194.html [April 2019].

Kobrin, J.L., Patterson, B.F., Shaw, E.J., Mattern, K.D., and Barbuti, S.M. (2008). *Validity of the SAT® for Predicting First-Year College Grade Point Average.* Research Report No. 2008-5. New York, NY: College Board.

Kosciw, J.G., Palmer, N.A., Kull, R.M., and Greytak, E.A. (2012). The effect of negative school climate on academic outcomes for LGBT youth and the role of in-school supports. *Journal of School Violence, 12*(1), 45-63.

Kostyo, S., Cardichon, J., and Darling-Hammond, L. (2018). *Making ESSA's Equity Promise Real: State Strategies to Close the Opportunity Gap.* Palo Alto, CA: Learning Policy Institute.

Kotok, S., Sakiko, I., and Bodovski, K. (2016). School climate and dropping out of school in the era of accountability. *American Journal of Education, 122*(4), 569-599.

Kraft, M.A. (2019).Teacher effects on complex cognitive skills and social-emotional competencies. *Journal of Human Resources, 54,* 1-36.

Kraus, N., Slater, J., Thompson, E.C., Hornickel, J., Strait, D.L., Nicol, R., and White-Schwoch, T. (2014). Music enrichment programs improve the neural encoding of speech in at-risk children. *Journal of Neuroscience, 34*(36), 11913-11918.

Kurlaender, M., Reardon, S., and Jackson, J. (2008). *Middle School Predictors of High School Achievement in Three California School Districts.* Santa Barbara, CA: California Dropout Research Project, University of California, Santa Barbara.

Kurlaender, M., and Yun, J.T. (2005). Fifty years after Brown: New evidence of the impact of school racial composition on student outcomes. *International Journal of Educational Policy, Research, and Practice: Reconceptualizing Childhood Studies, 6*(1), 51-78.

Kurlaender, M., and Yun, J.T. (2007). Measuring school racial composition and student outcomes in a multiracial society. *American Journal of Education, 113*(2), 213-242.

Kurz, A., Elliott, S.N., Lemons, C.J., Zigmond, N., Kloo, A., and Kettler, R.J. (2014). Assessing opportunity to learn for students with disabilities in general and special education classes. *Assessment for Effective Intervention, 40*(1), 24-39.

Ladd, H.F. (2017). No Child Left Behind: A deeply flawed federal policy. *Journal of Policy Analysis and Management, 36*(2), 461-469.

Ladd, H.F., and Sorensen, L.C. (2017). Returns to teacher experience: Student achievement and motivation in middle school. *Education Finance and Policy, 12*(2), 241-279.

Lafortune, J., Rothstein, J., and Whitmore-Schanzenbach, D. (2016). *School Finance Reform and the Distribution of Student Achievement.* NBER Working Paper No. 22011. Cambridge, MA: National Bureau of Economic Research.

Lauff, E., and Ingels, S.J. (2013). *Education Longitudinal Study of 2002 (ELS: 2002): A First Look at 2002 High School Sophomores 10 Years Later* (NCES 2014-363). Washington, DC: U.S. Department of Education, National Center for Education Statistics.

Lee, V.E., and Burkam, D.T. (2003). Dropping out of high school: The role of school organization and structure. *American Educational Research Journal, 40*(2), 353-393.

Leibowitz, A. (1977). Parental inputs and children's achievement. *The Journal of Human Resources, 12*(2), 242-251.

Levenstein, R. (2014). *2014 My Voice, My School Survey Quality Profile: Technical Report.* Chicago, IL: University of Chicago Consortium on School Research.

Lewis, A.E., and Diamond, J.B. (2015). *Despite the Best Intentions: How Racial Inequality Thrives in Good Schools.* Oxford, UK: Oxford University Press.

Lindsay, C.A., and Hart, C.M. (2017). Teacher race and school discipline. *Education Next, 17*(1). Available: https://www.educationnext.org/teacher-race-and-school-discipline-suspensions-research [April 2019].

Lipscomb, S., Haimson, J., Liu, A.Y., Burghardt, J., Johnson, D.R., and Thurlow, M.L. (2017). *Preparing for Life After High School: The Characteristics and Experiences of Youth in Special Education: Findings from NLTS 2012. Volume 1: Comparisons with Other Youth.* Washington, DC: U.S. Department of Education, Institute of Education Sciences, National Center for Education Evaluation and Regional Assistance. Available: https://ies.ed.gov/ncee/projects/evaluation/disabilities_nlts2012.asp [April 2019].

Losen, D.J., Martinez, T., and Gillespie, J. (2012). *Suspended Education in California.* Los Angeles, CA: The Civil Rights Project. Available: https://www.civilrightsproject.ucla.edu/resources/projects/center-for-civil-rights-remedies/school-to-prison-folder/summary-reports/suspended-education-in-california [April 2019].

Lynch, M., and Cicchetti, D. (1998). An ecological-transactional analysis of children and contexts: The longitudinal interplay among child maltreatment, community violence, and children's symptomatology. *Development and Psychopathology, 10*(2), 235-257.

MacNeil, A.J., Prater, D.L., and Busch, S. (2009). The effects of school culture and climate on school achievement. *International Journal of Leadership in Education, 12*(1), 73-84.

Madaus, G.F. (1988). The influence of testing on the curriculum. In L. Tanner (Ed.), *Critical Issues in Curriculum, NSSE Yearbook, Part 1* (pp. 83-121). Chicago, IL: University of Chicago Press.

Maehr, M.L., and Meyer, H.A. (1997). Understanding motivation and schooling: Where we've been, where we are, and where we need to go. *Educational Psychology Review, 9*(4), 371-409.

Magnuson, K. (2018). *Indicators of the Equity of School Readiness.* Paper prepared for the Committee on Developing Indicators of Educational Equity. Available: http://sites.nationalacademies.org/cs/groups/dbassesite/documents/webpage/dbasse_193236.pdf [May 2019].

Maloney, E.A., Ramirez, G., Gunderson, E.A., Levine, S.C., and Beilock, S.L. (2015). Intergenerational effects of parents' math anxiety on children's math achievement and anxiety. *Psychological Science, 26*(9), 1480-1488. Available: https://doi.org/10.1177/0956797615592630 [April 2019].

Manlove, J. (1998). The influence of high school dropout and school disengagement on the risk of school-age pregnancy. *Journal of Research on Adolescence, 8*(2), 187-220.

Manolitsis, G., Georgiou, G.K., and Tziraki, N. (2013). Examining the effects of home literacy and numeracy environment on early reading and math acquisition. *Early Childhood Research Quarterly, 28*(4), 692-703.

Manski, C.F., and Wise, D.A. (1983). *College Choice in America*. Cambridge, MA: Harvard University Press.

Maple, S.A., and Stage, F.K. (1991). Influences on the choice of math/science major by gender and ethnicity. *American Educational Research Journal, 28*(1), 37-60. Available: https://doi.org/10.3102/00028312028001037 [April 2019].

Margolin, G., and Gordis, E.B. (2000). The effects of family and community violence on children. *Annual Review of Psychology, 51*(1), 445-479.

Massachusetts Department of Elementary and Secondary Education. (2017). *Massachusetts Comprehensive Assessment System*. Available: http://www.doe.mass.edu/mcas/2017 questionnaire/faq.html [December 2018].

Massey, D.S., and Denton, N.A. (1988). The dimensions of residential segregation. *Social Forces*, 67:281-315

McCaffrey, D.F., Lockwood, J.R., Koretz, D.M., Louis, T.A., and Hamilton, L.S. (2004). Models for value-added modeling of teacher effects. *Journal of Educational and Behavioral Statistics, 29*(1), 67-102.

McCaffrey, D.F., Sass, T.R., Lockwood, J.R., and Mihaly, K. (2009). The intertemporal variability of teacher effect estimates. *Education Finance and Policy, 4*(4), 572-606.

McCarthy, K.F., Ondaatje, E.H., Zakaras, L., and Brooks, A. (2004). *Gifts of the Muse: Reframing the Debate about the Benefits of the Arts*. Santa Monica, CA: RAND Corporation. Available: https://www.rand.org/content/dam/rand/pubs/monographs/2005/RAND_MG218.pdf [April 2019].

McClelland, M.M., Morrison, F.J., and Holmes, D.L. (2000). Children at risk for early academic problems: The role of learning-related social skills. *Early Childhood Research Quarterly, 15*(3), 307-329. Available: http://www.sciencedirect.com/science/article/pii/S0885200600000697 [August 2017].

McEachin, A., Domina, T., and Penner, A.M. (2017). *Understanding the Effects of Middle School Algebra: A Regression Discontinuity Approach*. RAND Working Paper 1209 (Vol. October). Available: http://www.asdk12.org/forms/uploads/MSprogram.pdf [April 2019].

McFarland, J., Hussar, B., Wang, X., Zhang, J., Wang, K., Rathbun, A., Barmer, A., Forrest Cataldi, E., and Bullock Mann, F. (2018). *The Condition of Education 2018* (NCES 2018-144). Washington, DC: U.S. Department of Education, National Center for Education Statistics. Available: https://nces.ed.gov/pubsearch/pubsinfo.asp?pubid=2018144 [December 2018].

McLanahan S. (2009). Fragile families and the reproduction of poverty. *The Annals of the American Academy of Political and Social Science, 621*(1), 111-131.

McLanahan, S., Jeager, K., and Cantena, K. (2019). Children in fragile families. *Oxford Handbook of Children and Law*, J. Dwyer (Ed.). Oxford, UK: Oxford University Press. Available: https://doi.org/10.1093/oxfordhb/9780190694395.013.12.

McLanahan, S., Tach, L., and Schneider, D. (2013). The causal effects of father absence. *Annual Review of Sociology, 39*, 399-427. Available: https://doi.org/10.1146/annurev-soc-071312-145704.

Meece, J.L., Blumenfeld, P.C., and Hoyle, R.H. (1988). Students' goal orientations and cognitive engagement in classroom activities. *Journal of Educational Psychology, 80*(4), 514-523.

Mehrens, W.A. (1998). Consequences of assessment: What is the evidence? *Education Policy Analysis Archive, 6*(13). Available: http://olam.ed.asu.edu.epaa.v6n13.html [April 2019].

Mickelson, R.A. (2005). How tracking undermines race equity in desegregated schools. In J. Petrovich and A.S. Wells (Eds.), *Bringing Equity Back*. New York, NY: Teachers College Press.

Middlebrooks, J.S., and Audage, N.C. (2008). *The Effects of Childhood Stress on Health across the Lifespan*. Atlanta, GA: Centers for Disease Control and Prevention, National Center for Injury Prevention and Control. Available: http://health-equity.lib.umd. edu/932/1/Childhood_Stress.pdf [January 2019].

Mihaly, K., McCaffrey, D.F., Staiger, D.O., and Lockwood, J.R. (2013). *A Composite Estimator of Effective Teaching*. MET Project Research Paper, 1-51. Available: http:// k12education.gatesfoundation.org/download/?Num=2551&filename=MET_Composite_ Estimator_of_Effective_Teaching_Research_Paper.pdf [April 2019].

Miller, R.T., Murnane, R.J., and Willett, J.B. (2008). Do teacher absences impact student achievement? Longitudinal evidence from one urban school district. *Educational Evaluation and Policy Analysis, 30*(2), 181-200. Available: https://doi.org/10.3102/0162373708318019 [April 2019].

Müller, W., and Karle, W. (1993). Social selection in educational systems in Europe. *European Sociological Review, 9*(1), 1-23. Available: https://doi.org/10.1093/oxfordjournals.esr. a036652 [December 2018].

Mulligan, G.M., Hastedt, S., and McCarroll, J.C. (2012). *First-Time Kindergartners in 2010-11: First Findings from the Kindergarten Rounds of the Early Childhood Longitudinal Study, Kindergarten Class of 2010-11* (ECLS-K:2011) (NCES 2012-049). Washington, DC: U.S. Department of Education, National Center for Education Statistics. Available: https://eric.ed.gov/?id=ED533795 [April 2019].

Nagaoka, J. (2018). *Graduating from High School Ready for College*. Paper prepared for the Committee on Developing Indicators of Educational Equity. Available: http://sites. nationalacademies.org/cs/groups/dbassesite/documents/webpage/dbasse_193231.pdf [May 2019].

National Academy of Sciences, National Academy of Engineering, and Institute of Medicine. (2007). *Rising Above the Gathering Storm: Energizing and Employing America for a Brighter Economic Future*. Washington, DC: The National Academies Press. Available: https://doi.org/10.17226/11463 [April 2019].

National Academies of Sciences, Engineering, and Medicine. (2017). *Evaluation of the Achievement Levels for Mathematics and Reading on the National Assessment of Educational Progress*. Washington, DC: The National Academies Press. Available: https://doi. org/10.17226/23409 [April 2019].

National Center for Children in Poverty. (2019). *Child Poverty*. Available: http://www.nccp. org/topics/childpoverty.html [April 2019].

National Center for Education Statistics. (2012). *NAEP: Looking Ahead—Leading Assessments into the Future*. Washington, DC: Author.

National Commission on Excellence in Education. (1983). *A Nation at Risk: the Imperative for Education Reform*. A Report to the Nation and the Secretary of Education. Washington, DC: U.S. Government Printing Office. Available: https://www.edreform.com/ wp-content/uploads/2013/02/A_Nation_At_Risk_1983.pdf [August 2019].

National Council of Teachers of Mathematics. (2014). *Principles to Actions: Ensuring Mathematical Success for All*. Reston, VA: Author.

National Research Council. (1995). *Measuring Poverty: A New Approach*. Washington, DC: National Academy Press. Available: https://doi.org/10.17226/4759 [April 2019].

National Research Council. (1998). *Preventing Reading Difficulties in Young Children*. Washington, DC: National Academy Press. Available: https://doi.org/10.17226/6023 [April 2019].

National Research Council. (2009). *Protecting Student Records and Facilitating Education Research: A Workshop Summary*. M. Hilton, rapporteur. Washington, DC: The National Academies Press. Available: https://doi.org/10.17226/12514 [December 2018].

National Research Council. (2010). *Getting Value Out of Value-Added: Report of a Workshop*. Washington, DC: The National Academies Press. Available: https://doi.org/10.17226/12820 [April 2019].

National Research Council. (2011). *High School Dropout, Graduation, and Completion Rates: Better Data, Better Measures, Better Decisions*. Washington, DC: The National Academies Press. Available: https://doi.org/10.17226/13035 [April 2019].

National Research Council. (2012). *Key National Education Indicators: Workshop Summary*. Washington, DC: The National Academies Press. Available: https://doi.org/10.17226/13453 [April 2019].

National Research Council. (2013). *Monitoring Progress Toward Successful K–12 STEM Education: A Nation Advancing?* Washington, DC: The National Academies Press. Available: https://doi.org/10.17226/13509 [April 2019].

National Research Council and Institute of Medicine. (2000). *From Neurons to Neighborhoods: The Science of Early Childhood Development*. Washington, DC: The National Academies Press. Available: https://doi.org/10.17226/9824 [April 2019].

National Research Council and Institute of Medicine. (2004). *Engaging Schools: Fostering High School Students' Motivation to Learn*. Washington, DC: The National Academies Press. Available: https://doi.org/10.17226/10421 [April 2019].

National Scientific Council on the Developing Child. (2014). *Excessive Stress Disrupts the Architecture of the Developing Brain*. Working Paper 3. Cambridge, MA: Harvard University.

National Scientific Council on the Developing Child and National Forum on Early Childhood Policy and Programs. (2011). *Building the Brain's "Air Traffic Control" System: How Early Experiences Shape the Development of Executive Function*. Working Paper No. 11. Cambridge, MA: Center on the Developing Child, Harvard University.

National Urban League. (2019). *Standards of Equity and Excellence: A Lens on ESSA State Plans*. New York, NY: Author. Available: http://ncos.iamempowered.com/pdf/ESSA%20 Full%20Report.pdf [May 2019].

Neild, R.C., and Balfanz, R. (2006). *Unfulfilled Promise: The Dimensions and Characteristics of Unfulfilled Promise: The Dimensions and Characteristics of Philadelphia's Dropout Crisis, 2000-05*. Philadelphia, PA: Philadelphia Youth Transitions Collaborative.

New York City Department of Education. (2017). *NYC School Survey*. Available: http://schools.nyc.gov/Accountability/tools/survey/default.htm [December 2018].

Newmann, F.M., Bryk, A.S., and Nagaoka, J. (2001). *Authentic Intellectual Work and Standardized Tests: Conflict or Coexistence?* Chicago, IL: Consortium on Chicago School Research.

Newman, F.M., Wehlage, G.G. and Lamborn, S.D. (1992) The significance and sources of student engagement. In F.M. Newman (Ed.), *Student Engagement and Achievement in American Secondary Schools* (pp. 11-39). New York, NY: Teachers College Press.

Norbury, H., Wong, M., Wan, M., Reese, K., Dhillon, S., and Gerdeman, R. (2012). *Using the Freshman On-Track Indicator to Predict Graduation in Two Urban Districts in the Midwest Region*. (Issues & Answers Report, REL 2012–No.134). Washington, DC: U.S. Department of Education, Institute of Education Sciences, National Center for Education Evaluation and Regional Assistance, Regional Educational Laboratory Midwest. Available: https://files.eric.ed.gov/fulltext/ED531422.pdf [April 2019].

Nord, C., Roey, S., Perkins, R., Lyons, M., Lemanski, N., Brown, J., and Schuknecht, J. (2011). *Nation's Report Card: America's High School Graduates* (NCES 2011462). U.S. Department of Education, National Center for Education Statistics. Washington, DC: U.S. Government Printing Office.

Nye, B., Konstantopoulos, S., and Hedges, L.V. (2004). How large are teacher effects? *Educational Evaluation and Policy Analysis, 26*(3), 237-257. Available: https://doi.org/10.3102/01623737026003237 [April 2019].

Oakes, J. (1985). *Keeping Track: How Schools Structure Inequality.* New Haven, CT: Yale University Press.

Oakes, J. (1986). *Education Indicators: A Guide for Policy Makers.* Santa Monica, CA: RAND Corporation.

Oakes, J. (1990). Chapter 3: Opportunities, achievement, and choice: Women and minority students in science and mathematics. *Review of Research in Education, 16*(1), 153-222. Available: https://doi.org/10.3102/0091732X016001153 [April 2019].

O'Malley, M., Voight, A., Renshaw, T., and Eklund, K. (2014). School climate, family structure, and academic achievement: A study of motivation effects. *School Psychology Quarterly, 30*(1), 360-377.

Oreopoulos, P., and Salvanes, K.G. (2011). Priceless: The nonpecuniary benefits of schooling. *Journal of Economic Perspectives, 25*(1), 159-184.

Orfield, G. (2013). Housing segregation produces unequal schools causes and solutions. In P.L. Carter and K.G. Welner (Eds.), *Closing the Opportunity Gap* (pp. 40-60). Oxford, UK: Oxford University Press.

Orfield, G., Ee, J., Frankenberg, E., and Siegel-Hawley, G. (2016). *Brown at 62: School Segregation by Race, Poverty and State.* Los Angeles, CA: Civil Rights Project.

Orfield, G., and Lee, C. (2005). *Why Segregation Matters: Poverty and Educational Inequality.* Cambridge, MA: The Civil Rights Project.

Orfield, G., Siegel-Hawley, G., and Kucsera, J. (2014). *Sorting Out Deepening Confusion on Segregation Trends.* Cambridge, MA: The Civil Rights Project.

Osher, D., Cantor, P., Berg, J., Strayer, L., and Rose, T. (2018). Drivers of human development: How relationships and context shape learning and development. *Applied Developmental Science.* Available: https://doi.org/10.1080/10888691.2017.1398650 [May 2019].

Owens, A. (2016). Inequality in children's contexts: The economic segregation of households with and without children. *American Sociological Review, 81*(3), 549-574.

Owens, A., Reardon, S.F., and Jencks, C. (2016). Income segregation between schools and school districts. *American Educational Research Journal, 53*(4), 1159-1197.

Pallais, A., and Turner, S. (2006). Opportunities for low-income students at top colleges and universities: Policy initiatives and the distribution of students. *National Tax Journal, 59*(2), 357-386.

Papay, J.P., and Kraft, M.A. (2015). Productivity returns to experience in the teacher labor market: Methodological challenges and new evidence on long-term career improvement. *Journal of Public Economics, 130,* 105-119.

Pattillo, M. (2013). *Black Picket Fences: Privilege and Peril among the Black Middle Class.* Chicago, IL: University of Chicago Press.

Peisner Feinberg, E.S., Burchinal, M.R., Clifford, R.M., Culkin, M.L., Howes, C., Kagan, S.L. and Yazejian, N. (2001), The relation of preschool child care quality to children's cognitive and social developmental trajectories through second grade. *Child Development, 72*(5), 1534-1553.

The Pell Institute for the Study of Equal Opportunity in Higher Education. (2017). *Indicators of Higher Education Equity in the United States: 2017 Historical Trend Report.* Philadelphia, PA: University of Pennsylvania.

Penner, A.M., Domina, T., Penner, E.K., and Conley, A.M. (2015). Curricular policy as a collective effects problem: A distributional approach. *Social Science Research, 52,* 627-641. Available: https://doi.org/10.1016/j.ssresearch.2015.03.008 [April 2019].

Persico, C., Figlio, D., and Roth, J. (2016). *Inequality before Birth: The Developmental Consequences of Environmental Toxicants.* NBER Working Paper No. w22263. Cambridge, MA: National Bureau of Economic Research.

Persson, M. (2015). Classroom climate and political learning: Findings from a Swedish panel study and comparative data. *Political Psychology, 36*(5), 587-601.

Pettigrew, T.F., and Tropp, L.R. (2006). A meta-analytic test of intergroup contact theory. *Journal of Personality and Social Psychology, 90*(5), 751-783. Available: http://dx.doi.org/10.1037/0022-3514.90.5.751 [April 2019].

Phillips, D.A., Gormley, W.T., and Lowenstein, A.E. (2009). Inside the pre-kindergarten door: Classroom climate and instructional time allocation in Tulsa's pre-K programs. *Early Childhood Research Quarterly, 24*(3), 213-228. Available: http://dx.doi.org/10.1016/j.ecresq.2009.05.002 [April 2019].

Phillips, M. (2011). Parenting, time use, and disparities in academic outcomes. In G.J. Duncan and R. Murnane (Eds.), *Whither Opportunity?: Rising Inequality, Schools, and Children's Life Chances.* New York, NY: Russell Sage Foundation.

Pianta, R.C., Belsky, J., Houts, R., Morrison, F., and National Institute of Child Health and Human Development Early Child Care Research Network. (2007). Teaching. Opportunities to learn in America's elementary classrooms. *Science, 315*(5820), 1795-1796.

Plank, S., DeLuca, S., and Estacion, A. (2008). High school dropouts and the role of career and technical education: A survival analysis of surviving high school. *Sociology of Education, 81*(4), 345-370.

Planty, M., and Carlson, D. (2010). *Understanding Education Indicators: A Practical Primer for Research and Policy.* New York, NY: Teachers College Press.

Pleis, J.R., and Lucas, J.W. (2009). Summary health statistics for U.S. adults: National Health Interview Survey, 2007. *Vital and Health Statistics 10*(240), 1-159. Available: http://www.cdc.gov/nchs/data/series/sr_10/sr10_240.pdf [April 2019].

Podlozny, A. (2000). Strengthening verbal skills through the use of classroom drama: A clear link. *Journal of Aesthetic Education, 34*(3-4), 239-276.

Powell, D.R., Son, S.H., File, N., and San Juan, R.R. (2010). Parent-school relationships and children's academic and social outcomes in public school pre-kindergarten. *Journal of School Psychology, 48*(4), 269-292.

Pre-Kindergarten Task Force. (2017). *The Current State of Scientific Knowledge on Pre-Kindergarten Effects.* Washington, DC: Brookings Institution and the Duke Center for Child and Family Policy. Available: https://www.brookings.edu/research/puzzling-it-out-the-current-state-of-scientific-knowledge-on-pre-kindergarten-effects/ [August 2019].

Pritzker, J.B., Bradach, J.L., and Kaufmann, K. (2015). *Achieving Kindergarten Readiness for All Our Children: A Funder's Guide to Early Childhood Development from Birth to Five.* Chicago: J.B. and M.K. Pritzker Family Foundation and the Bridgespan Group.

Putnam, R.D. (2001). *Bowling Alone: The Collapse and Revival of American Community.* New York, NY: Simon & Schuster.

Putnam, R.D. (2015). *Our Kids: The American Dream in Crisis.* New York, NY: Simon and Schuster.

Pynoos, R.S., Steinberg, A.M., Layne, C.M., Liang, L.J., Vivrette, R.L., Briggs, E. C., Kisiel, C., Habib, M., Belin, T.R., and Fairbank, J. (2014). Modeling constellations of trauma exposure in the National Child Traumatic Stress Network Core Data Set. *Psychological Trauma: Theory, Research, Practice, and Policy, 6*(Suppl. 1), S9-S17.

Rahman, T., Fox, M.A., Ikoma, S., and Gray, L. (2017). *Certification Status and Experience of U.S. Public School Teachers: Variations across Student Subgroups* (NCES 2017-056). U.S. Department of Education, National Center for Education Statistics. Available: https://nces.ed.gov/pubs2017/2017056.pdf [May 2019].

Ramey, G., and Ramey, V.A. (2010). The rug rat race. *Brookings Papers on Economic Activity, Spring,* 129-176.

Raver, C.C. (2004). Placing emotional self regulation in sociocultural and socioeconomic contexts. *Child Development, 75*(2), 346-353.

Raver, C.C., Gershoff, E.T., and Aber, J.L. (2007). Testing equivalence of mediating models of income, parenting, and school readiness for White, Black, and Hispanic children in a national sample. *Child Development, 78*(1), 96-115.

Raver, C.C., Smith-Donald, R., Hayes, T., and Jones, S.M. (2005). *Self-Regulation across Differing Risk and Sociocultural Contexts: Preliminary Findings from the Chicago School Readiness Project.* Biennial meeting of the Society for Research in Child Development, Atlanta, GA.

Ready, D.D. (2010). Socioeconomic disadvantage, school attendance, and early cognitive development: The differential effects of school exposure. *Sociology of Education, 83*(4), 271-286.

Ready, D.D. (2017). *School-Based Predictors of Student ELA and Mathematics Development.* Paper prepared for the Committee on Developing Indicators of Educational Equity.

Reardon, S.F. (2011). The widening academic achievement gap between the rich and the poor: New evidence and possible explanations. In G.J. Duncan and R.J. Murnane (Eds.), *Whither Opportunity? Rising Inequality, Schools, and Children's Life Chances.* New York, NY: Russell Sage Foundation.

Reardon, S.F. (2016). School segregation and racial academic achievement gaps. *RSF: The Russell Sage Foundation Journal of the Social Sciences, 2*(5), 34-57.

Reardon, S.F., and Bischoff, K. (2011). Income inequality and income segregation. *American Journal of Sociology, 116*(4), 1092-153.

Reardon, S.F., Bischoff, K., Owens, A., and Townsend, J.B. (2018). Has income segregation really increased? Bias and bias correction in sample-based segregation estimates. *Demography, 55*(6), 2129-2160. Available: https://doi.org/10.1007/s13524-018-0721-4 [April 2019].

Reardon, S.F., Fox, L. and Townsend, J. (2015). Neighborhood income composition by race and income, 1990 2009. *The Annals of the American Academy of Political and Social Science, 660,* 78-97.

Reardon, S.F., Kalogrides, D., and Shores, K. (2019). The geography of racial/ethnic test score gaps. *The American Journal of Sociology, 124*(4). Available: https://www.journals.uchicago.edu/doi/full/10.1086/700678 [May 2019].

Reardon, S.F., and Owens, A. (2014). 60 years after Brown: Trends and consequences of school segregation. *Annual Review of Sociology, 40,* 199-218.

Reardon, S.F., and Yun, J.T. (2001). Suburban racial change and suburban school segregation, 1987–1995. *Sociology of Education, 74*(2), 79-101.

Ribar, D.C. (2015). Why marriage matters for child well-being. *Future of Children, 25*(2), 11-27.

Richburg-Hayes, L. (2018). *Immediate Transitions: Common Pathways after High School Graduation.* Paper prepared for the Committee on Developing Indicators of Educational Equity. Available: http://sites.nationalacademies.org/cs/groups/dbassesite/documents/webpage/dbasse_193234.pdf [May 2019].

Richters, J.E., and Martinez, P. (2016). The NIMH community violence project: I. Children as victims of and witnesses to violence. *Psychiatry, 56*(1), 7-21.

Rice, J.K. (2013). Learning from experience: Evidence on the impact and distribution of teacher experience and the implications for teacher policy. *Education Finance and Policy,* *8*(3), 332-348.

Riegle-Crumb, C. (2006). The path through math: Course sequences and academic perfor-mance at the intersection of race-ethnicity and gender. *American Journal of Education,* *113*(1), 101-122.

Riegle-Crumb, C., and Humphries, M. (2012). Exploring bias in math teachers' perceptions of students' ability by gender and race/ethnicity. *Gender & Society, 26*(2), 290-322.

Rimm-Kaufman, S.E., Curby, T.W., Grimm, K.J., Nathanson, L., and Brock, L.L. (2009). The contribution of children's self-regulation and classroom quality to children's adaptive behaviors in the kindergarten classroom. *Developmental Psychology, 45*(4), 958-972.

Rivkin, S.G., Hanushek, E.A., and Kain, J.F. (2005). Teachers, schools, and academic achieve-ment. *Econometrica, 73*(2), 417-458.

Rockoff, J.E. (2004). The impact of individual teachers on student achievement: Evidence from panel data. *The American Economic Review, 94*(2), 247-252.

Roderick, M., Coca, V., and Nagaoka, J. (2011). Potholes on the road to college: High school effects in shaping urban students' participation in college application, four-year college enrollment, and college Match. *Sociology of Education, 84*(3), 178-211.

Roderick, M., Nagaoka, J., and Allensworth, E. (2006). *From High School to the Future: A First Look at Chicago Public School Graduates' College Enrollment, College Prepa-ration, and Graduation from Four-Year Colleges.* UChicago Consortium on Chicago School Research. Available: https://consortium.uchicago.edu/publications/high-school-future-first-look-chicago-public-school-graduates-college-enrollment [May 2019].

Roderick, M., Nagaoka, J., Coca, V., and Moeller, E. (2008). *From High School to the Future: Potholes on the Road to College. Research Report.* UChicago Consortium on Chicago School Research. Available: https://consortium.uchicago.edu/publications/high-school-future-potholes-road-college [May 2019].

Romero, M., and Lee, Y-S. (2007). *A National Portrait of Chronic Absenteeism in the Early Grades.* New York, NY: National Center for Children in Poverty, Mailman School of Public Health at Columbia University. Available: http://www.nccp.org/publications/pdf/text_771.pdf [April 2019].

Rosenkranz, T., de la Torre, M., Stevens, W.D., and Allensworth, E.M. (2014). *Free to Fail or On-Track to College: Why Grades Drop When Students Enter High School and What Adults Can Do About It.* Chicago: UChicago Consortium on Chicago School Research.

Rothstein, J.M. (2004). College performance predictions and the SAT. *Journal of Econometrics, 121*(1), 297-317.

Rothstein, J.M. (2017). Measuring the impacts of teachers: Comment. *American Economic Review, 107*(6), 1656-1684.

Rothstein, R. (2015). The racial achievement gap, segregated schools, and segregated neigh-borhoods: A constitutional insult. *Race and Social Problems, 7*(1), 21-30. Available: http://dx.doi.org/10.1007/s12552-014-9134-1 [April 2019].

Rothwell, J.T., and Massey, D.S. (2010). Density zoning and class segregation in U.S. metro-politan areas. *Social Science Quarterly, 91*(5), 1123-1143.

Rumberger, R.W. (1995). Dropping out of middle school: A multilevel analysis of students and schools. *American Educational Research Journal, 32*(3), 583-625.

Rumberger, R.W. (2011). *Dropping Out: Why Students Drop Out of High School and What Can Be Done About It.* Cambridge, MA: Harvard University Press. Available: http://dx.doi.org/10.4159/harvard.9780674063167 [April 2019].

Rumberger, R.W., and Losen, D.H. (2016). *The High Cost of Harsh Discipline and Its Disparate Impact.* Los Angeles, CA: The Civil Rights Project. Available: https://www. civilrightsproject.ucla.edu/resources/projects/center-for-civil-rights-remedies/school-to-prison-folder/federal-reports/the-high-cost-of-harsh-discipline-and-its-disparate-impact [April 2019].

Rumberger, R.W., and Palardy, G.J. (2005). Test scores, dropout rates, and transfer rates as alternative indicators of high school performance. *American Educational Research Journal, 41*(1), 3-42.

Rumberger, R.W., and Plasman, J. (2018). *Developing Equity Indicators for On-Time Graduation.* Paper prepared for the Committee on Developing Indicators of Educational Equity. Available: http://sites.nationalacademies.org/cs/groups/dbassesite/documents/webpage/dbasse_193232.pdf [May 2019].

Runningen, A. (2014). *Meeting the Challenges of Adolescent Psychotherapy: Review.* Available: https://www.questia.com/library/journal/1P3-3681363721/meeting-the-challenges-of-adolescent-psychotherapy [April 2019].

Russell Sage Foundation. (1912). *A Comparative Study of Public School Systems in the Forty-Eight States.* New York, NY: Author.

Russell, V.J., Ainley, M., and Frydenberg, E. (2005). Student motivation and engagement. *Schooling Issues Digest.* Australian Government, Department of Education, Science and Training.

Ryan, C.L., and Bauman, K. (2016). *Educational Attainment in the United States: 2015.* Washington, DC: United States Census Bureau. Available: https://www.census.gov/content/dam/Census/library/publications/2016/demo/p20-578.pdf [April 2019].

Sabol, T.J., and Pianta, R.C. (2017). The state of young children in the United States: School readiness. In E. Votruba-Drzal, and E. Dearing (Eds.), *Handbook of Early Childhood Development Programs, Practices, and Policies* (pp. 1-17). Wiley Blackwell Handbooks of Developmental Psychology. Hoboken, NJ: Wiley-Blackwell.

Sadler, P.M., Sonnert, G., Hazari, Z., and Tai, R.H. (2012). Stability and volatility of STEM career interest in high school: A gender study. *Science Education, 96*(3), 411-427.

Sadler, P.M., Sonnert, G., Coyle, H.P., Cook-Smith, N., and Miller, J.L. (2013). The influence of teachers' knowledge on student learning in middle school physical science classrooms. *American Educational Research Journal, 50*(5), 1020-1049. Available: https://doi.org/10.3102/0002831213477680 [April 2019].

Sanders, W.L., and Rivers, J.C. (1996). *Cumulative and Residual Effects of Teachers on Future Student Academic Achievement* (Research Progress Report). Knoxville, TX: University of Tennessee Value-Added Research and Assessment Center.

Saporito, S., and Sohoni, D. (2007). Mapping educational inequality: Concentrations of poverty among poor and minority students in public schools. *Social Forces, 85*(3), 1227-1253.

Sartain, L., Allensworth, E.M., and Porter, S. (2015). *Suspending Chicago's Students: Differences in Discipline Practices across Schools.* CCSR Research Report. UChicago Consortium on Chicago School Research. Available: https://consortium.uchicago.edu/publications/suspending-chicagos-students-differences-discipline-practices-across-schools [April 2019].

Sayer, L.C., Bianchi, S.M., and Robinson, J.P. (2004a). Are parents investing less in children? Trends in mothers' and fathers' time with children. *American Journal of Sociology, 110*(1), 1-43.

Scarborough, H.S., and Dobrich, W. (1994). On the efficacy of reading to preschoolers. *Developmental Review, 14*(3), 245-302.

Schanzenbach, D.W., Nunn, R., Bauer, L., Mumford, M., and Breitwieser, A. (2016). *Seven Facts on Noncognitive Skills from Education to the Labor Market.* Washington, DC: The Hamilton Project. Available: http://www.hamiltonproject.org/assets/files/seven_facts_noncognitive_skills_education_labor_market.pdf [December 2018].

Schwartz, H. (2010). *Housing Policy Is School Policy: Economically Integrative Housing Promotes Academic Success in Montgomery County, Maryland.* New York, NY: The Century Foundation.

Sebastian, J., and Allensworth, E. (2012). The influence of principal leadership on classroom instruction and student learning: A study of mediated pathways to learning. *Educational Administration Quarterly, 48*(4), 626-663.

Sebastian, J., Allensworth, E., and Stevens, W.D. (2014). The influence of school leadership on classroom participation: Examining configurations of organizational supports. *Teachers College Record, 116*(8), 1-36.

Sebastian, J., Huang, H., and Allensworth, E. (2017). Examining integrated leadership systems in high schools: Connecting principal and teacher leadership to organizational processes and student outcomes. *School Effectiveness and School Improvement, 28*(3), 463-488.

Shanahan, T., and Lonigan, C.J. (2010). The National Early Literacy Panel: A summary of the process and the report. *Educational Researcher, 39*(4), 279-285. Available: https://doi.org/10.3102/0013189X10369172 [April 2019].

Sharkey, P. (2010). The acute effect of local homicides on children's cognitive performance. *Proceedings of the National Academy of Sciences of the United States of America, 107*(26), 11733-11738.

Sharkey, P. (2014). Spatial segmentation and the Black middle class. *American Journal of Sociology, 119*(4), 903-954.

Sharkey, P., Schwartz, A.E., Ellen, I.G., and Lacoe, J. (2014). High stakes in the classroom, high stakes on the street: The effects of community violence on student's standardized test performance. *Sociological Science, 1*, 199-220. doi: 10.15195/v1.a14.

Sharkey, P.T., Tirado-Strayer, N., Papachristos, A.V., and Raver, C.C. (2012). The effect of local violence on children's attention and impulse control. *American Journal of Public Health, 102*(12), 2287-2293.

Shavelson, R.J., McDonnell, L.M., Oakes, J., Carey, N.B., and Picus, L. (1987). *Indicator Systems for Monitoring Science and Mathematics Education.* Santa Monica, CA: RAND Corporation. Available: https://www.rand.org/pubs/reports/R3570.html [December 2018].

Shavit, Y., and Blossfeld, H-P. (Eds.). (1993). *Persistent Inequality: Changing Educational Attainment in Thirteen Countries. Social Inequality Series.* Boulder, CO: Westview Press.

Shonkoff, J.P. (2010). Building a new biodevelopmental framework to guide the future of early childhood policy. *Child Development, 81*(1), 357-367. doi: 10.1111/j.1467-8624.2009.01399.x.

Siegel-Hawley, G. (2012). *How Non-Minority Students Also Benefit from Racially Diverse Schools.* Research Brief No. 8. Washington, DC: The National Coalition on School Diversity. Available: https://school-diversity.org/pdf/DiversityResearchBriefNo8.pdf [April 2019].

Sigle-Rushton, W., and McLanahan, S. (2002). For richer or poorer? Marriage as an anti-poverty strategy in the United States. *Population, 57*(3), 509-528.

Sigle-Rushton, W., and McLanahan, S. (2004). Father absence and child well-being. In D. Moynihan, T. Speeding, and L. Rainwater (Eds.), *The Future of the Family* (pp. 116-155). New York, NY: Russell Sage Foundation.

Simzar, R., Domina, T., and Tran, C. (2016). Eighth-grade algebra course placement and student motivation for mathematics. *AERA Open, 2*(1), 233285841562522.

Skiba, R.J., Chung, C.G., Trachok, M., Baker, T.L., Sheya, A., and Hughes, R.L. (2014). Parsing disciplinary disproportionality: Contributions of infraction, student, and school characteristics to out-of-school suspension and expulsion. *American Educational Research Journal, 51*, 640-670.

Skinner, E., and Belmont, M. (1993). Motivation in the classroom: Reciprocal effects of teacher behavior and student engagement across the school year. *Journal of Educational Psychology, 85*(4), 571-581

Skinner, E.A., Kindermann, T.A., Connell, J.P., and Wellborn, J.G. (2009). Engagement and disaffection as organizational constructs in the dynamics of motivational development. In K.R. Wenzel and A. Wigfield (Eds.), *Educational Psychology Handbook Series. Handbook of Motivation at School* (pp. 223-245). New York, NY: Routledge/Taylor & Francis Group.

Smerdon, B.A. (1999). Engagement and achievement: Differences between African-American and white high school students. *Research in the Sociology of Education and Socialization, 12*(1), 103-134.

Smerillo, N.E., Reynolds, A.J., Temple, J.A., and Ou, S.-R. (2018). Chronic absence, eighth-grade achievement, and high school attainment in the Chicago Longitudinal Study. *Journal of School Psychology, 67*, 163-178.

Smith, M.S. (1988). Educational indicators. *The Phi Delta Kappan, 69*(7), 487-491.

Snyder, T.D., de Brey, C., and Dillow, S.A. (2016). *Digest of Education Statistics 2014* (NCES 2016-006). National Center for Education Statistics, Institute of Education Sciences, U.S. Department of Education. Available: https://nces.ed.gov/pubs2016/2016006.pdf [April 2019].

Snyder, T.D., de Brey, C., and Dillow, S.A. (2019). *Digest of Education Statistics 2017* (NCES 2018-070). National Center for Education Statistics, Institute of Education Sciences, U.S. Department of Education. Available: https://nces.ed.gov/pubs2018/2018070.pdf [April 2019].

Sorensen, L., Fox, A., Jung, H., and Martin, E.G. (2019). Lead exposure and academic achievement: Evidence from childhood lead poisoning prevention efforts. *Journal of Population Economics, 32*(1), 179-218. Available: https://doi.org/10.1007/s00148-018-0707-y [April 2019].

Stecher, B.M., Holtzman, D.J., Garet, M.S., Hamilton, L., Engberg, J., Steiner, E.D., Robyn, A., Baird, M.D., Gutierrez, I.A., Peet, E.D., Brodziak de los Reyes, I., Fronberg, K., Weinberger, G., Hunter, G.P., and Chambers, J. (2018). *Improving Teaching Effectiveness. Final Report: The Intensive Partnerships for Effective Teaching through 2015-2016.* Santa Monica, CA: RAND Corporation.

Stein, M.K., and Lane, S. (1996). Instructional tasks and the development of student capacity to think and reason: An analysis of the relationship between teaching and learning in a reform mathematics project. *Educational Research and Evaluation, 2*(1), 50-80.

Steinberg, L.D., Brown, B.B., and Dornbusch, S.M. (1996). *Beyond the Classroom: Why School Reform Has Failed and What Parents Need to Do.* New York, NY: Simon & Schuster.

Steinberg, M.P., Allensworth, E., and Johnson, D.W. (2015). What conditions support safety in public schools? The influence of school organizational practices on student and teacher reports of safety in Chicago Public Schools. Pp. 118-131 in D.J. Rosen (Ed.), *Closing the School Discipline Gap: Equitable Remedies for Excessive Exclusion.* New York, NY: Teachers College Press.

Storch, S., and Whitehurst, G. (2001). The role of family and home in the literacy development of children from low-income backgrounds. *New Directions in Child and Adolescent Development, 92*, 53-71.

Strive Together. (2013). *Cradle to Career Core Outcome Areas.* Available: https://www. strivetogether.org/wp-content/uploads/2017/06/StriveTogether-cradle-to-career-outcomes-research-publication.pdf [April 2019].

Strive Together. (2018). *Theory of Action.* Available: https://www.strivetogether.org/wp-content/uploads/2018/10/Theory-of-Action-Poster.pdf [April 2019].

Stuit, D., O'Cummings, M., Norbury, H., Heppen, J., Dhillon, S., Lindsay, J., and Zhu, B. (2016). *Identifying Early Warning Indicators in Three Ohio School Districts* (REL 2016–118). Washington, DC: U.S. Department of Education, Institute of Education Sciences, National Center for Education Evaluation and Regional Assistance, Regional Educational Laboratory Midwest. Available: http://ies.ed.gov/ncee/edlabs/projects/project.asp?ProjectID=358 [April 2019].

Sum, A., Khatiwada, I., McLaughlin, J., and Palma, S. (2009). *The Consequences of Dropping Out of High School: Joblessness and Jailing for High School Dropouts and the High Cost for Taxpayers.* Boston, MA: Center for Labor Market Studies at Northeastern University.

Susperreguy, M.I., and Davis-Kean, P. (2016). Maternal math talk in the home and math skills in preschool children. *Early Education and Development, 27*(6), 841-857. Available: https://www.researchgate.net/publication/299374572_Maternal_math_talk_in_the_home_and_math_skills_in_preschool_children [April 2019].

Taylor, J.J., Buckley, K., Hamilton, L.S., Stecher, B.M., Read, L., and Schweig, J. (2018). *Choosing and Using SEL Competency Assessments: What Schools and Districts Need to Know.* Santa Monica, CA: RAND Corporation. Available: http://measuringsel.casel. org/pdf/Choosing-and-Using-SEL-Competency-Assessments_What-Schools-and-Districts-Need-to-Know.pdf [December 2018].

Taylor, L.L. (2018). *Real Needs and Real Resources: Identifying Indicators of School Funding Equity.* Paper prepared for the Committee on Developing Indicators of Educational Equity. Available: http://sites.nationalacademies.org/cs/groups/dbassesite/documents/webpage/dbasse_193235.pdf [May 2019].

Taylor, R.D., Oberle, E., Durlak, J.A., and Weissberg, R.P. (2017). Promoting positive youth development through school-based social and emotional learning interventions: A meta-analysis of follow-up effects. *Child Development, 88*(4), 1156-1171.

Thapa, A., Cohen, J., Higgins-D'Alessandro, A., and Guffey, S. (2012). *School Climate Research Summary: August 2012.* School Climate Brief, Number 3. New York, NY: National School Climate Center.

Thapa, A., Cohen, J., Guffey, S., and Higgins-D'Alessandro, A. (2013). A review of school climate research. *Review of Educational Research, 83*, 357-385. Available: https://doi.org/10.3102/0034654313483907 [August 2019].

Thomas, A., and Sawhill, I. (2005). For love and money: The impact of family structure on family income. *Future of Children, 15*(2), 57-74.

Thompson, K.D. (2017). What blocks the gate? Exploring current and former English learners' math course-taking in secondary school. *American Educational Research Journal, 54*(4), 757-798.

Tierney, A.T., Krizman, J., and Kraus N. (2015). Music training alters the course of adolescent auditory development. *Proceedings of the National Academy of Sciences of the United States, 112*(32), 10062-10067. Available: https://www.pnas.org/content/112/32/10062 [April 2019].

Torney-Purta, J. (2002). The school's role in developing civic engagement: A study of adolescents in twenty-eight countries. *Applied Developmental Science 6*(4), 203-312.

Tout, K., Starr, R., Soli, M., Moodie, S., Kirby, G., and Boller, K. (2010). *Compendium of Quality Rating Systems and Evaluations.* Administration for Children and Families. Available: http://www.acf.hhs.gov/programs/opre/cc/childcare_quality/compendium_qrs/qrs_compendium_final.pdf [April 2019].

Truman, H.S. (1947). *235. Statement by the President Making Public a Report of the Commission on Higher Education.* Available: https://trumanlibrary.org/publicpapers/index. php?pid=1852 [April 2019].

Trusty, J. (2002). Effects of high school course-taking and other variables on choice of science and mathematics college majors. *Journal of Counseling & Development, 80*(4), 464-474. Available: https://doi.org/10.1002/j.1556-6678.2002.tb00213.x [April 2019].

Tschannen-Moran, M., Parish, J., and DiPaola, M. (2006). School climate: The interplay between interpersonal relationships and student achievement. *Journal of School Leadership, 16*(4), 386.

Turner, J.C., Thorpe, P.K., and Meyer, D.K. (1998). Students' reports of motivation and negative affect: A theoretical and empirical analysis. *Journal of Educational Psychology, 90*, 758-771.

Umansky, I.M. (2016). Leveled and exclusionary tracking: English learners' access to academic content in middle school. *American Educational Research Journal, 53*(6), 1792-1833.

U.S. Department of Education. (2013). *For Each and Every Child—A Strategy for Education Equity and Excellence.* Washington, DC: Author. Available: https://www2.ed.gov/about/ bdscomm/list/eec/equity-excellence-commission-report.pdf [April 2019].

U.S. Department of Education. (2017). *Nevada Department of Education Consolidated State Plan under the Every Student Succeeds Act.* Available: http://www.doe.nv.gov/ upload-edFiles/ndedoenvgov/content/Boards_Commissions_ Councils/ESSA_Adv_Group/ESSA_ Nevada_Consolidated_State_ Plan_4.3.17_Finalrev.pdf [December 2018].

U.S. Department of Education, Office of Civil Rights. (2014). *Civil Rights Data Collection: Data Snapshot: Teacher Equity.* Washington, DC: Author. Available: https://ocrdata. ed.gov/Downloads/CRDC-Teacher-Equity-Snapshot.pdf [April 2019].

U.S. Department of Education, Office of Civil Rights. (2016). *2013-2014 Civil Rights Data Collection: A First Look.* Washington, DC: Author. Available: https://www2.ed.gov/ about/offices/list/ocr/docs/2013-14-first-look.pdf [August 2018].

U.S. Department of Health and Human Services, Administration for Children and Families. (2010). *Head Start Impact Study. Final Report.* Washington, DC: Author.

U.S. Government Accountability Office. (2009). *Student Achievement: Schools Use Multiple Strategies to Help Students Meet Academic Standards, Especially Schools with Higher Proportions of Low-Income and Minority Students.* GAO-10-18. Washington, DC: Author. Available: https://www.gao.gov/assets/300/298455.pdf [April 2019].

U.S. Government Accountability Office. (2011). *Key Indicator Systems: Experiences of Other National and Subnational Systems Offer Insights for the United States.* Washington, DC: Author. Available: http://www.gao.gov/products/GAO-11-396 [April 2017].

U.S. Government Accountability Office. (2016). *Better Use of Information Could Help Agencies Identify Disparities and Address Racial Discrimination.* Washington, DC: Author.

U.S. Government Accountability Office. (2018). *K–12 EDUCATION: Discipline Disparities for Black Students, Boys, and Students with Disabilities.* GAO 18-258. Washington, DC: Author. Available: https://www.gao.gov/products/GAO-18-258 [December 2018].

Utah Education Policy Center. (2012). *Research Brief: Chronic Absenteeism.* Salt Lake City, UT: The University of Utah. Available: https://daqy2hvnfszx3.cloudfront.net/wp-content/ uploads/sites/2/2017/05/23104652/ChronicAbsenteeismResearchBrief.pdf [April 2019].

Van der Kolk, B.A. (2005). Developmental Trauma Disorder: Toward a rational diagnosis for children with complex trauma histories. *Psychiatric Annals, 35*(5), 401-408.

Vandell, D.L., Belsky, J., Burchinal, M., Steinberg, L., Vandergrift, N., and NICHD Early Child Care Research Network. (2010). Do effects of early child care extend to age 15 years? *Child Development, 81*(3), 737-756. Available: https://doi.org/10.1111/j.1467-8624.2010.01431.x.

Vernon-Feagans, L., Garrett-Peters, P.T., Willoughby, M.T., and Mills-Koonce, W.R. (2012). The family life project key investigators: Chaos, poverty, and parenting: predictors of early language development. *Early Childhood Research Quarterly, 27*(3), 339-351.

Wagner, C. (2017). *School Segregation Then & Now: How to Move toward a More Perfect Union.* Alexandria, VA: The Center for Public Education.

Waldfogel, J. (2012). The role of out-of-school factors in the literacy problem. *Future of Children, 22*(2), 39-54. Available: https://eric.ed.gov/?id=EJ996186 [April 2019].

Walker, E., Tabone, C., and Weltsek, G. (2011). When achievement data meet drama and arts integration. *Language Arts, 88*(5), 365-372. Available: https://eric.ed.gov/?id=EJ923719 [April 2019].

Wang, M.-T., and Eccles, J.S. (2013). School context, achievement motivation, and academic engagement: A longitudinal study of school engagement using a multidimensional perspective. *Learning and Instruction, 28,* 12-23. Available: http://dx.doi.org/10.1016/j.learninstruc.2013.04.002 [April 2019].

Wang, W., Vaillancourt, T., Brittain, H.L., Mcdougall, P., Krygsman, A., Smith, D.W., Cunningham, C.E., Haltigan, J.D., and Hymel, S. (2014). School climate, peer victimization, and academic achievement: Results from a multi-informant study. *School Psychology Quarterly, 29*(3), 360-377.

Wang, X. (2013). Why students choose STEM majors: Motivation, high school learning, and postsecondary context of support. *American Educational Research Journal, 50*(5), 1081-1121. Available: https://www.insidehighered.com/sites/default/server_files/files/Wang%20AERJ%20Oct%202013.pdf [April 2019].

Wang, Y., and Benner, A.D. (2014). Parent-child discrepancies in educational expectations: Differential effects of actual versus perceived discrepancies. *Child Development, 85*(3), 891-900.

Ware, N.C., and Lee, V.E. (1988). Sex differences in choice of college science majors. *American Educational Research Journal, 25*(4), 593-614. Available: https://doi.org/10.3102/00028312025004593 [April 2019].

Washbrook, E.V., and Waldfogel, J. (2011). *On Your Marks: Measuring the School Readiness of Children in Low-to-Middle Income Families.* Bristol, UK: Resolution Foundation.

Weiland, C., and Yoshikawa, H. (2013). Impacts of a prekindergarten program on children's mathematics, language, literacy, executive function, and emotional skills. *Child Development, 84*(6), 2112-2130.

Weissberg, R.P., Durlak, J.A., Domitrovich, C.E., and Gullotta, T.P. (2015). Social and emotional learning: Past, present, and future. In J.A. Durlak, C.E. Domitrovich, R.P. Weissberg, T.P. Gullotta, and J. Comer, (Eds.), *Handbook of Social and Emotional Learning: Research and Practice* (pp. 3-19). New York, NY: Guilford Press.

Wells, A.S., Fox, L., and Cordova-Cobo, D. (2016). *How Racially Diverse Schools and Classrooms Can Benefit All Students.* New York, NY: The Century Foundation. Available: https://production-tcf.imgix.net/app/uploads/2016/02/09142501/HowRaciallyDiverse_AmyStuartWells-11.pdf [April 2019].

West, J., Denton, K., and Germino Hausken, E. (2000). *America's Kindergartners* (NCES 2000–070). Washington, DC: Government Printing Office, U.S. Department of Education, National Center for Education Statistics.

Willoughby, T.M., Kupersmidt, J., and Voegler-Lee, M.E. (2011). Is preschool executive function causally related to academic achievement? *Child Neuropsychology, 18*(1), 79-91. Available: https://www.researchgate.net/publication/51251790_Is_Preschool_Executive_Function_Causally_Related_to_Academic_Achievement [April 2019].

Winner, E., Goldstein, T., and Vincent-Lancrin, S. (2013). *Art for Art's Sake? The Impact of Arts Education.* Paris, France: OECD Publishing. Available: http://www.oecd.org/education/ceri/arts.htm [April 2019].

Wolf, S., Magnuson, K.A., and Kimbro, R.T. (2017). Family poverty and neighborhood poverty: Links with children's school readiness before and after the Great Recession. *Children and Youth Services Review, 79*, 368-384. Available: http://dx.doi.org/10.1016/j.childyouth.2017.06.040 [April 2019].

Wong, M.D., Shapiro, M.F., Boscardin, W.J., and Ettner, S.L. (2002). Contribution of major diseases to disparities in mortality. *The New England Journal of Medicine, 347*(20), 1585-1592. Available: https://www.ncbi.nlm.nih.gov/pubmed/12432046 [April 2019].

Workman, S., and Ullrich, R. (2017). Quality 101: Identifying the core components of a high-quality early childhood program. *Early Childhood.* Washington, DC: Center for American Progress. Available: https://www.americanprogress.org/issues/early-childhood/reports/2017/02/13/414939/quality-101-identifying-the-core-components-of-a-high-quality-early-childhood-program [December 2018].

Yazzie-Mintz, E. (2007). *Voices of Students on Engagement: A Report on the 2006 High School Survey of Student Engagement.* Bloomington, IN: Indiana University.

Yen, W.M. (2010). *Measuring Student Growth with Large-Scale Assessments in an Education Accountability System.* Princeton, NJ: Educational Testing Service. Available: https://www.ets.org/Media/Research/pdf/YenReactorSession1.pdf [December 2018].

Yoshikawa, H., Weiland, C., Brooks-Gunn, J., Burchinal, M., Espinosa, L., Gormley, W.T., Ludwig, J., Magnuson, K., Phillips, D., and Zaslow, M. (2013). *Investing in Our Future: The Evidence Base on Preschool.* Washington, DC: Society for Research in Child Development.

Yudof, M.G., Levin, B., Moran, R., Ryan, J.E., and Bowman, K.L. (2011). *Educational Policy and the Law,* 5th Edition. Boston, MA: Cengage Learning.

Zau, A.C., and Betts, J.R. (2008). *Predicting Success, Preventing Failure: An Investigation of the California High School Exit Exam.* San Francisco, CA: Public Policy Institute of California. Available: https://www.ppic.org/content/pubs/report/R_608AZR.pdf [April 2019].

Zhang, A., Musu-Gillette, L., and Oudekerk, B.A. (2016). *Indicators of School Crime and Safety: 2015* (NCES 2016-079/NCJ 249758). Washington, DC: National Center for Education Statistics, U.S. Department of Education, and Bureau of Justice Statistics, Office of Justice Programs, U.S. Department of Justice.

Zill, N., and West, J. (2001). *Entering Kindergarten: A Portrait of American Children When They Begin School. Findings from the Condition of Education, 2000.* Washington, DC: National Center for Education Statistics.

Appendix A

Review of Existing Data Systems

As part of its information gathering, the committee investigated the potential usefulness of existing data systems for monitoring progress (or lack thereof) toward equity among groups of children enrolled in K–12 education. The first part of this appendix provides a brief historical overview of the interest in education indicator systems in the United States. The second part describes and assesses the relevance for educational equity of the major existing data systems that regularly monitor the state of education. Box A-1 lists the criteria that informed the committee's assessment.

Other information from the committee's work is in the next two appendices: Appendix B reviews existing publications of indicators that are potentially relevant for monitoring K–12 educational equity, and Appendix C summarizes the data and methodological challenges in implementing the committee's recommended indicators.

A BRIEF HISTORY OF EDUCATION INDICATORS[1]

1840-1960

Interest in developing a system of education indicators in the United States began in the mid-19th century. On the part of the federal government, the constitutionally mandated decennial census as early as 1840

---

[1]This section draws heavily on materials on the National Center for Education Statistics website, including *Federal Education Data Collection—Celebrating 150 Years*. Available: https://nces.ed.gov/surveys/annualreports/pdf/Fed_Ed_Data_Collection_Celebrating_150_Years.pdf.

BOX A-1
Criteria for Assessing Data Systems for
Educational Equity Indicators

1. Published on a regular, frequent basis—at least annually.
2. Available for subnational geographic areas, including states, school districts, and, ideally, schools or school attendance areas, as appropriate.
3. High-quality when assessed on measures of nonsampling error (e.g., accurate reporting of student enrollment) and on measures of sampling error (for survey-based data).
4. Available for groups of children of interest for education equity (see Chapter 2 text), as defined by race and ethnicity, gender, family income (or equivalent measure of socioeconomic resources), disability status, immigrant status, and English language capability.
   a. For immigrant children, indicative of time of entry into the United States to appropriately include/exclude them in equity indicators (e.g., exclude from a high school graduation measure if they arrived only a year before graduation).
   b. For English-language learners, when possible, indicative of the number of years spent in an English-learner program, whether a student waived out of English-learner instruction, and time and type of reclassification to English-proficient status.
5. Measures contextual factors, such as neighborhood income and family type composition for student groups of interest (see Chapter 3).[a]
6. Measures students' educational outcomes for student groups of interest in three domains comprising seven indicators, each with one or more constructs to be measured (see Chapter 4):
   Domain A: Kindergarten readiness
      Indicator 1: Disparities in academic readiness (reading/literacy, numeracy/math skills)
      Indicator 2: Disparities in self-regulation and attention skills
   Domain B: K–12 learning and engagement (measured at multiple levels/grades)
      Indicator 3: Disparities in engagement in schooling (attendance/absenteeism, academic engagement)
      Indicator 4: Disparities in performance in coursework (success in classes, accumulating credits to be on track to graduate, grades/GPA)
      Indicator 5: Disparities in performance on tests (reading/math/science achievement, learning growth in reading/math/science achievement)

asked about education and learning: the 1840-1930 censuses asked about literacy (for people over age 20); the 1850 through 2000 censuses asked about school attendance; and the 1940-2000 censuses asked about educational attainment (Citro, 2012). The school attendance and educational attainment questions are now part of the monthly American Community

Domain C: Educational attainment
Indicator 6: Disparities in on-time high school graduation
Indicator 7: Disparities in postsecondary readiness (enrollment in college, entry into the workforce, enlistment in the military)
7. Measures school-provided opportunities to learn for student groups of interest in four domains comprising nine indicators, each with one or more constructs to be measured (see Chapter 5):
Domain D: Extent of racial, ethnic, and economic segregation
Indicator 8: Disparities in students' exposure to racial, ethnic, and economic segregation (concentrated poverty in schools, racial segregation within and across schools)
Domain E: Equitable access to high-quality early learning programs
Indicator 9: Disparities in access to and participation in high-quality pre-K programs (availability and participation in licensed pre-K programs)
Domain F: Equitable access to high-quality curricula and instruction
Indicator 10: Disparities in access to effective teaching (teachers' years of experience, teachers' credentials/certification, racial and ethnic diversity of the teaching force)
Indicator 11: Disparities in access to and enrollment in rigorous coursework (availability/enrollment in advanced, rigorous coursework, availability/enrollment in Advanced Placement, International Baccalaureate, and dual enrollment programs, availability/enrollment in gifted and talented programs)
Indicator 12: Disparities in curricular breadth (availability/enrollment in coursework in the arts, social sciences, sciences, and technology)
Indicator 13: Disparities in access to high-quality academic supports (access to and participation in formalized systems of tutoring or other types of academic supports including special education services and services for English learners)
Domain G: Equitable access to supportive school and classroom environments
Indicator 14: Disparities in school climates (perceptions of safety, academic support, academically focused culture, teacher-student trust)
Indicator 15: Disparities in nonexclusionary discipline practices (out-of-school suspensions/expulsions)
Indicator 16: Disparities in nonacademic supports for student success (supports for emotional, behavioral, mental, physical health)

---

[a]Although we do not propose indicators of context, they would be critical to inform efforts of school systems to work with other sectors to combat root causes of poverty and other factors that adversely affect students' educational attainment.

Survey (ACS), which began in 2005 and collects information on a broad range of topics (see below).

The 1840 census had questions about schools and school enrollment. In 1867, Congress, recognizing the need for and interest in greater detail about public education, such as school finances, teachers, and graduates, chartered a national Department of Education to "[collect] such statistics and facts as shall show the condition and progress of education in the several States" (P.L. 39-73, 14 Stat. 434). In the 1890 Second Morrill Act, Congress required the collection of similar statistics for private K–12 education.

Congress abolished the new department in 1869 but not the statistics function, which it vested in a Bureau of Education in the Department of the Interior. The bureau (renamed the Office of Education) was transferred to the Federal Security Agency in 1939 and to the new U.S. Department of Health, Education, and Welfare in 1953—becoming the National Center for Education Statistics (NCES) in 1962. It was made part of the new Department of Education in 1980 and is now located in the department's Institute of Education Sciences.

Nongovernmental organizations also early became involved in reporting on the state of education. In 1912, the Russell Sage Foundation published a report that ranked states according to such indicators as school attendance and school expenditures (Russell Sage Foundation, 1912).

Federal education statistics originally focused on characteristics of school districts, such as counts of students and teachers and revenue and expenditure information. The Annual Report of the Commissioner of Education provided comprehensive data on city school systems from 1871 to 1918; the Biannual Survey of Education in the United States became the vehicle for reporting such information from 1917 to 1955. The last biannual report expanded coverage to include suburban and rural school systems. In addition, the Office of Education published a series of annual studies on current expenditures per pupil in city school systems from 1918 through 1960.

1960-Present

In the 1960s, the Office of Education suspended the collection of general local school system data for several years to meet the data collection needs set forth in recently enacted legislation, especially the 1958 National Defense Education Act and the 1965 Elementary and Secondary Education Act. The need to examine local school systems in their entirety rather than solely in terms of program segments, as well as demands by the educational community for basic data, led to the resumption of school district and school-based data collection in 1967. That work was carried out by NCES, initially on a sample basis through the Elementary & Secondary Education General Information Survey (ELSEGIS) and now as a census through the Common Core of Data (see below).

In terms of measuring student outcomes, including achievement levels at different grades, and satisfactory completion of high school and preparation for adult success, the federal government took some steps in the late 1960s and early 1970s in this direction. These included initiation of the National

Assessment of Educational Progress (NAEP) in 1969 by NCES, under the guidance of an advisory group (now the statutorily authorized National Assessment Governing Board, NAGB), which measures student achievement at several grades (see below). It also included the first of NCES's longitudinal studies, the National Longitudinal Study of the High School Class of 1972, which follow students and their achievements over time (see below).

These initiatives, however, did not represent a sustained effort to monitor student outcomes, let alone educational equity, by relating outcomes to education resources for states, school districts, and schools. Publication of *A Nation at Risk* (National Commission on Excellence in Education, 1983) represented a milestone in national attention to education and is widely credited with stimulating an earnest and sustained push for close monitoring of student achievement and implementation of education reforms to raise achievement levels (Bryk and Hermanson, 1993; Ginsburg, Noell, and Plisko, 1988). It was followed in 1984 by the Department of Education's "*Wall Chart*," which, while informative, brought attention to the limitations of the available data and spurred interest in developing ways to enable valid state-by-state comparisons of student achievement (Ginsburg, Noell, and Plisko, 1988). Soon afterwards, there was a push for increased sampling for NAEP that would enable reporting of state-level achievement data, and in 2001 Congress required that all states participate in NAEP's reading and math assessments for grades 4 and 8 every 2 years as a condition of receiving Title I funds under the Elementary and Secondary Education Act. NAEP also added the Trial Urban District Assessment, which allows large cities to monitor trends and compare their students' achievements with those of other cities.

In 2002, Title II of the Educational Technical Assistance Act provided for grants to states to develop and expand the Statewide Longitudinal Data Systems. These systems are intended to pull together administrative data on students and follow their progress through at least K–12 (see below). Some systems follow students from preschool through college and entry into the workforce, and some systems include links to students' teachers. Grants, which are managed by NCES, began in 2005 and have been made to all but three states.

Parallel with data collection efforts under NCES, other offices in the U.S. Department of Education have mandated data collection for administration of federal education funding programs to states and school districts and for enforcement of civil rights law. The Office of Civil Rights in the department, for example, collects useful data (see below).

Individual school districts, particularly in large cities, have developed their own sets of indicators with which to monitor student progress and achievement, not only in terms of test scores, but also on other dimensions, such as absences. Many indicators are produced according to agreed-on definitions and standards developed by the Department of Education working with state education agencies. Some key indicators, however, such as grade-specific achievement tests, vary across states, and other indicators are tailored to the needs of the particular district.

Researchers working with individual school districts have conducted surveys, obtained administrative records, and constructed measures that have potential use for a system of educational equity indicators. The discussion of the committee's proposed indicators in Chapters 4 and 5 references studies that suggest the value and feasibility of collecting relevant information, even though these studies have not themselves generated data with the breadth of geographic and demographic detail needed for a comprehensive indicator system. The Stanford Education Data Archive (SEDA) is an exception, in part (see below). It assembles achievement test scores for race and ethnicity groups and grades from all school districts in the nation and links test scores to characteristics of the school districts, such as urban/suburban/rural, obtained from publicly available sources.

RELEVANT DATA SYSTEMS

This section describes and assesses data systems that have at least some of the information that would be required for an accurate, informative report on educational equity in U.S. K–12 education. Some systems are based on surveys, while others are based on administrative records or both kinds of data. The assessment covers the following data systems, using the criteria listed in Box A-1 (above):

- Data systems that provide geographic and demographic detail:

    o American Community Survey (ACS)
    o Civil Rights Data Collection (CRDC)
    o Common Core of Data (CCD)
    o National Assessment of Educational Progress (NAEP)
    o Small-Area Income and Poverty Estimates (SAIPE) Program
    o Stanford Education Data Archive (SEDA)
    o Statewide Longitudinal Data Systems (SLDS)

- Data systems that provide national information and demographic detail:

    o Annual Social and Economic Supplement and School Enrollment Supplement of the Current Population Survey (CPS) (CPS ASEC and CPS SES, respectively)
    o NCES Longitudinal Surveys

- o   NCES Household Education Survey (NHES)
- o   NCES National Teacher and Principal Survey (NTPS)

American Community Survey[2]

The ACS is a large survey of the U.S. population, covering about 300,000 households per month and about 3.6 million households per year. Conducted by the U.S. Census Bureau, it became operational in 2005 and includes content that was previously on the "long-form" questionnaire that was part of the decennial census.[3]

Assessed against the criteria for measuring education equity, the ACS performs as follows:

Frequency and geographic detail—publishes data annually, including 1-year aggregations for the preceding calendar year for areas of at least 65,000 population and 5-year aggregations for school districts, census tracts, and block groups (5-year aggregates are necessary to provide reliable estimates for small areas).

Data quality—good quality in terms of low unit and item nonresponse rates and comparisons with other surveys; sampling error is low for larger geographic areas, but becomes large for small geographic areas.

Student groups of interest—collects data on age, gender, race and ethnicity, income, selected disabilities, immigrant status and year of immigration, and language spoken at home of students enrolled in public and private schools.

Contextual factors—collects data on a wide variety of characteristics that can be tabulated for geographic areas, giving the population composition by race and ethnicity, income and poverty status, family type, and other attributes.

Educational outcomes—has information on college enrollment and/ or employment of college-age young people, which is relevant to Domain C (educational attainment), but post-high school status cannot be tied to the responsible school district or school.

Educational opportunities—has information on the composition of the student body (in terms of income, race and ethnicity, and other characteristics) in a school district for public and private schools, which is relevant to Domain D (extent of segregation), but mea-

---

[2]For information on the ACS, see https://www.census.gov/programs-surveys/acs/; see also relevant articles in Anderson, Citro, and Salvo (2012).

[3]As of 2010, the census includes a limited set of questions to meet its constitutional mandate to provide information for congressional reapportionment and legislative redistricting.

sures of racial and socioeconomic segregation are not available for individual schools within a district.

For its Education Demographic and Geographic Estimates (EDGE) program (see below), NCES regularly commissions the Census Bureau to prepare detailed school district tabulations, using 5-year ACS data, to describe the distribution of school-age children and their parents by such characteristics as family income and race and ethnicity.[4] SEDA includes EDGE tabulations in its program (see below). It could also be possible to generate specialized tables for school districts, school attendance areas, and households of students attending a given school, given appropriate address information.[5] Sample size, however, limits the amount of geographic or substantive detail that the ACS can provide with sufficient reliability for use, although modeling can help (see SAIPE Program below).

Civil Rights Data Collection[6]

The federal government began collecting information in 1968 from school districts and schools with which to monitor and enforce laws prohibiting discrimination on the basis of race, gender, national origin, and disability in K–12 education (see Chapter 2). The CRDC program, known originally as the Elementary and Secondary School Survey, collects information biannually from school districts and schools, including juvenile justice facilities, charter schools, alternative schools, and schools serving only students with disabilities. The CRDC originally collected data from large samples of districts and schools; beginning with the 2011-2012 school year, the CRDC is a census of all districts and schools.

The CRDC is a mandatory data collection, authorized under the statutes and regulations implementing Title VI of the Civil Rights Act of 1964, Title IX of the Education Amendments of 1972, and Section 504 of the Rehabilitation Act of 1973, and under the Department of Education Organization Act (20 U.S.C. § 3413).[7]

The CRDC is a survey of all public schools and school districts in the United States. The program collects an extensive array of information, including student characteristics relevant to discrimination law, school and teacher characteristics, and school financial information. The CRDC database, with hundreds of data elements, is fully accessible to the public.

---

[4]See https://nces.ed.gov/programs/edge/Demographic/ACS; the latest available EDGE tables are for 2012-2016.

[5]Such work would require access to ACS microdata in a secure Federal Statistical Research Data Center; see https://www.census.gov/fsrdc.

[6]For information on the CRDC, see https://ocrdata.ed.gov.

[7]For details, see 34 CFR 100.6(b); 34 CFR 106.71; and 34 CFR 104.61.

School districts self-report and certify all their data. Assessed against our criteria for measuring education equity, the CRDC performs as follows:

Frequency and geographic detail—collects data biannually and makes them available about 2 years after collection; for the past four cycles has covered all school districts and schools (previously, collected information from a large sample); items collected on enrollment (in Part 1) are reported as of October 1 of the school year; items collected on students participating in Advanced Placement (AP) exams (in Part 2) are reported at the end of the school year.

Data quality—high, given that the Office of Civil Rights (OCR) in the Department of Education uses the data for enforcement and actively reviews the data and follows up on discrepancies.

Student groups of interest—collects data on numbers of students by race, gender, disability status (14 categories identified in the 1975 Individuals with Disabilities Education Act [IDEA] and students eligible only under Section 504 of the 1973 Rehabilitation Act), and limited English proficiency status; many variables are reported separately by these characteristics; has no information on student family income.

Contextual factors—has no information.

Educational outcomes—for student groups of interest, collects:
- o   information relevant to Domain B (K–12 learning and engagement), including Indicators 3 (engagement in schooling) and 4 (performance in course work), but not 5 (performance on tests).
- o   no information relevant to Domains A or C.

Educational opportunities—for student groups of interest, collects:
- o   some information relevant to Domain D (extent of racial, ethnic, and economic segregation), such as the percentage of students in various race/ethnicity groups, though not for income (also collects financial information that could be used to construct indicators of resources relative to numbers of disadvantaged children, but not including children in low-income families).
- o   basic information relevant to Domain E (high-quality early learning programs).
- o   extensive information relevant to Domain F (high-quality curricula and instruction), including Indicators 10 (effective teaching), 11 (access to and enrollment in rigorous coursework), and 12 (curricular breadth). Also ascertains the number of school counselors, which is relevant to Indicator 13 (access to high-quality academic interventions and supports).
- o   information that could potentially be used for Domain G (supportive school and classroom environments), such as inference as to discipline practices from information on disciplinary actions.

Common Core of Data[8]

The CCD is a long-standing program of NCES to provide basic statistics about K–12 education—its immediate predecessor was ELSEGIS, which began in 1967 on a sample survey basis. The CCD covers all public elementary and secondary schools and districts in the nation. Districts report school and district nonfinancial data to state educational agencies, which in turn submit the data through the U.S. Department of Education's ED*Facts* system,[9] according to definitions and reporting standards developed by NCES in cooperation with the Council of Chief State School Officers (CCSSO). School districts and state educational agencies submit financial data to the Census Bureau, on forms and definitions developed by the Bureau. Both ED*Facts* and Census Bureau data are regularly reviewed for accuracy and corrected as needed; data files become available with a 1- to 2-year lag.

Assessed against the criteria for measuring education equity, the CCD performs as follows:

Frequency and geographic detail—collects data annually from a census of school districts and schools and makes them available with about a 1-year lag; provides data for all levels of geography.

Data quality—high; NCES and the CCSSO have worked to ensure common definitions of enrollment, 4-year high school graduation rates (that take account that students starting high school in one school district may transfer to another), and other variables that could be subject to different interpretations; similarly, the Census Bureau has worked to establish common definitions for financial reporting.

Student groups of interest—obtains enrollment by grade by race and ethnicity, gender, and eligibility for free and reduced-price school lunches as a proxy for family income; obtains district-level counts of students with disabilities (total, not by type of disability) and English-language learners.

Contextual factors—has no information.

Educational outcomes—obtains limited information relevant to Domain C, Indicator 6 (on-time high school graduation), including district-level counts of high school diploma recipients and other

---

[8]For information on the CCD, see https://nces.ed.gov/ccd/. The Common Core of Data should not be confused with the Common Core State Standards (see http://www.corestandards.org/about-the-standards/).

[9]For information about ED*Facts*, see https://www2.ed.gov/about/inits/ed/edfacts/index.html, which contains a link to all of the ED*Facts* file specifications. In addition to the CCD, states satisfy the reporting requirements of the 2015 Every Student Succeeds Act through ED*Facts*.

high school completers and district-level counts of high school dropouts (access to data on grade, race and ethnicity, and gender composition of high school dropouts is restricted).

Educational opportunities—obtains limited information relevant to Domain F (high-quality curricula and instruction), such as training and length of service of teachers (relevant to Indicator 10), and whether high school AP courses are offered (relevant to Indicator 11); financial information submitted to the Census Bureau by districts and states breaks out sources of revenue, types of expenditures, and state and federal government support to districts by program (e.g., bilingual education) and could be used to construct indicators of resources relative to numbers of disadvantaged children at the district and state level, which is relevant to Domains E, F, and G.

Some points to note with regard to the nonfinancial information in the CCD include:

- The data do not identify immigrant children or U.S.-born children living with immigrant parents.
- The variable used for socioeconomic status—namely, children eligible for free and reduced-price school lunches—is increasingly less useful for this purpose (see discussion of Indicator 8, in Appendix C). The NCES commissioner has an initiative to develop a better measure of family income for children attending a school to address this problem (see discussion of the Small-Area Income and Poverty Estimates Program, below).
- Data for some groups, including children with disabilities and English-language learners, are not available for individual schools; this is also the case for counts of dropouts and high school completers.

National Assessment of Educational Progress[10]

NAEP is a long-standing and highly respected source of comparable data across the nation on student achievement at several grade levels in reading, math, and other subjects. Planning for NAEP began in 1964, and the first assessments were conducted in 1969 on a trial basis. The assessments, which were administered in public and private schools, covered citizenship, science, and writing performance of 17-year-old students in

---

[10]For information about NAEP, see https://nces.ed.gov/nationsreportcard/—particularly "History and Innovation" under "About" on the main page; see also Box 6-1 in Chapter 6 for a brief history of NAEP.

spring 1969 and of 9- and 13-year-old students and out-of-school 17-year-olds in fall 1969. Beginning in 1971 for reading and 1973 for math, NAEP has assessed 9-, 13-, and 17-year-olds every 4 years in what is termed "long-term trend NAEP," in which content has been kept as comparable as possible over time. Beginning in1990 for math and 1992 for reading, "main NAEP," in which content is modified about every 10 years to reflect changes in school curricula, has assessed 4th, 8th, and 12th graders.

The main NAEP sample size increased beginning in 1990 to support reliable results for states. Participation by students has always been voluntary, but in 2001 Congress required states to participate in the main NAEP 4th- and 8th-grade reading and math assessments as a condition of receiving Title I funding under the Elementary and Secondary Education Act. Beginning in 2002 Congress provided funding for selected urban school districts to participate in main NAEP as part of the Trial Urban District Assessment—eligible school districts must be above a specified number of enrolled students, a specified percentage of black or Hispanic students, and a specified percentage of students eligible for free and reduced-price school lunches.

Assessed against the criteria for measuring education equity, main NAEP in its current form performs as follows:

> *Frequency and geographic detail*—frequency varies by subject area and geographic level: every 2 years for math and reading for grades 4 and 8 for the nation, states, and selected urban districts (27 as of 2017); every 4 years for math and reading for grade 12 for the nation; periodically for science and writing for grades 4 and 8 for the nation, states, and selected urban districts; periodically for other subjects—technology and engineering literacy, arts, civics, geography, economics, and U.S. history—for the nation. The national assessments include public and private schools; the additional samples for states and selected urban districts include only public schools.
>
> *Data quality*—high, given the extensive methodological research that goes into constructing the content in each NAEP assessment and the sample design; response rates for schools in the period 2003-2015 were 95 percent or higher for grades 4 and 8 and 90-95 percent for grade 12, while student participation rates in the same period were several points lower than the school rates for grades 4 and 8 and considerably lower than the school rates for grade 12.[11]

---

[11]See *Focus on NAEP*, Figures 1 and 2; available: https://www.nationsreportcard.gov/focus_on_naep/files/g12_companion.pdf.

Student groups of interest—collects information on participating students' gender and race and ethnicity; also collects data on participation in the free and reduced-price school lunch program and disability and English-language-learner status (NAEP accommodates the latter two groups of students to encourage their participation).

Contextual factors—has no information.

Educational outcomes—estimates for each subject area the percentage of students achieving at specific proficiency levels, which is relevant to Domain B, Indicator 5 (performance on tests).

Educational opportunities—has no information.

Although main NAEP cannot be used to provide indicators for most school districts or for any schools, it is nonetheless important for an educational equity indicator system. Specifically, it can be used to calibrate state achievement test results of schools and school districts and thereby achieve greater comparability across states (see discussion of SEDA, below).

Small-Area Income and Poverty Estimates Program[12]

The U.S. Census Bureau runs the SAIPE Program, which began producing estimates in 1993 of the total number of school-age children and the number of school-age children in poverty by state, county, and school district for allocation of Title I funds under the Elementary and Secondary Education Act. The state and county estimates use a combination of 1-year ACS estimates and information from individual tax returns from the Internal Revenue Service (IRS) and records from the Supplemental Nutrition Assistance Program (SNAP, formerly food stamps), together with hierarchical Bayes modeling techniques to enhance the reliability of the estimates. The estimates for school districts use IRS data to estimate school-age poverty, adjusting those estimates to agree with the county estimates. This methodology currently produces just a single indicator of relevance for educational equity—namely, the concentration of poor school-age children by school district, which is relevant to Domain D, Indicator 8. The SAIPE estimates are released annually, about 1 year after completion of data collection.

The NCES commissioner has made it a priority to work with the Census Bureau to develop models with the ACS and administrative records to estimate poverty for students attending particular schools.[13] If successful, such estimates would be an improvement over the current reliance on the

---

[12]For information about SAIPE, see https://www.census.gov/programs-surveys/saipe.html.

[13]See http://magazine.amstat.org/blog/2018/10/01/meet-james-woodworth-nces-commissioner/.

number or percentage of children participating in free and reduced-price school lunch as an indicator of low-income status. Such models also could be extended to produce estimates for other groups of students of interest.

Stanford Education Data Archive[14]

SEDA, which is supported by the Institute of Education Sciences at the U.S. Department of Education and a number of foundations, has the following mission:

> [to harness] data to help scholars, policymakers, educators, and parents learn how to improve educational opportunity for all children. SEDA includes a range of detailed data on educational conditions, contexts, and outcomes in school districts and counties across the United States. It includes measures of academic achievement and achievement gaps for school districts and counties, as well as district-level measures of racial and socioeconomic composition, racial and socioeconomic segregation patterns, and other features of the schooling system.

SEDA performs on our criteria for useful data systems for monitoring educational equity as follows:

Frequency and geographic detail—currently (version 2.1) contains achievement data for the 2008-2009 through 2014-2015 school years for grades 3-8, obtained from state submissions for schools to ED*Facts*; school district student body composition from NCES's EDGE program of estimates derived from the 2006-2010 ACS; and financial and nonfinancial information on schools and districts from the CCD. It was scheduled to be updated in March 2019 to include data through the 2015-2016 school year and data broken down by economic status. It produces aggregate estimates for the nation, states, counties, and school districts. It also aggregates data for schools but does not release them as such.

Data quality—reflects the quality of the original data source; staff put in substantial effort to standardize data where possible—for example, by calibrating state achievement test scores to NAEP scores and generating constructed estimates that are as comparable as possible across states.

Student groups of interest—has information on gender, race and ethnicity, and disadvantaged status of students in each school grade (with disadvantaged status defined by the school district and generally

---

[14]For information about SEDA, see https://cepa.stanford.edu/seda/overview.

based on participation in the free and reduced-price school lunch program), but no information on disability, English-language-learner, or immigrant status.

Contextual factors—has the socioeconomic and demographic composition of school-age children in the relevant grades who attend public schools in a district, from the ACS-based EDGE.

Educational outcomes—has student test scores for students in grades 3-8 in reading and math and a number of constructed variables, such as calibrations with NAEP and estimates of student progress (e.g., scores of students in grade x in year t compared with the students in grade x+1 in year t+1); these variables are relevant for Domain B, Indicator 5 (performance on achievement tests).

Educational opportunities—has data from the CCD, which is relevant to some indicators in Domains F and G.

SEDA is a valuable resource for research on educational equity and has generated attention-getting research on inequitable educational outcomes and opportunities. All data aggregates are publicly available and appropriately protected for confidentiality (e.g., by adding small amounts of statistical noise). At present, however, SEDA could not support a full-fledged educational equity indicator system, such as we recommend, for several reasons. It is not up to date and does not as yet have a regular update schedule; it does not cover high school achievement or high school student and school characteristics, principally because there is not sufficient commonality among states as to when they test high school students; and it does not include important student characteristics attached to test scores—specifically, a good measure of family income or socioeconomic status or any measure of disability, immigrant, or English-language learning status.

Statewide Longitudinal Data Systems[15]

Interest in the SLDS began in the 1990s, when individual states began to recognize the value of linking administrative data on students, teachers, and schools to track student progress, identify correlates and perhaps causes of progress (or decline), and, more generally, to obtain a clearer and fuller picture of their K–12 educational systems. Some states sought to expand their databases to include postsecondary outcomes, such as employment and college enrollment, or to link their databases with other kinds

---

[15]For information about the SLDS, see https://nces.ed.gov/programs/SLDS/.

of state records, such as public assistance records. Often, states worked in cooperation with university education research centers.[16]

The federal government gave an important boost to these efforts when Congress provided funding for grants to states, administered by NCES, for the development and enhancement of statewide systems that linked student and other information over time using unique student identifiers. Funding was provided through Title II of the 2002 Educational Technical Assistance Act, with the first grants to states made in 2005. As of 2018, 47 states, American Samoa, the District of Columbia, Puerto Rico, and the Virgin Islands have received grants. Grants in 2005 and 2007 covered construction of data for K–12; grants in subsequent years have also supported linkages with pre-K, postsecondary, workforce, and teacher-student data. Because the awards are grants and not contracts, states have considerable latitude in the content and construction of an SLDS, although they must include variables that are required to be reported to the U.S. Department of Education, such as those in the CRDC.

The committee did not attempt to evaluate each state's SLDS for use in an educational equity indicator system, especially because access to the data is under the control of each state according to its own privacy policies and its interpretation of the Family Educational Rights and Privacy Act of 1974. To the extent that the organization(s) established to develop and operate a nationwide educational equity indicator system (see Chapter 6) can gain ready access to each state's SLDS, it is likely that the data would be highly useful for the purpose. Even without nationwide access, researchers' use of particular states' SLDS could generate ideas for new and refined indicators that could be produced from other, more readily available sources.

Other Data Collection Programs with Nationwide Detail

We briefly mention five other data collection programs that can serve such functions as providing input for national headline indicators (e.g., high school graduation rates for different student groups of interest) or supporting research that could lead to new or more refined indicators: the Annual Social and Economic Supplement and the School Enrollment Supplement of the CPS, NCES's longitudinal surveys, NCES's National Household Education Survey, and NCES's National Teacher and Principal Survey.

The CPS ASEC and CPS SES are household surveys conducted annually (the CPS ASEC in February-April and the CPS SES in October) as supple-

---

[16]For descriptions of Florida's and North Carolina's longitudinal database construction programs, see National Research Council (2009).

ments to the monthly CPS.[17] The CPS ASEC samples 100,000 households and obtains detailed information on employment and income over the preceding calendar year, disability and health status, language spoken at home, citizenship, when came to the United States, educational enrollment and attainment, veteran status, marital status, and family composition. State estimates are possible by averaging over 3 years. The CPS SES, which is supported by NCES, routinely gathers data on school enrollment and educational attainment for elementary, secondary, and postsecondary education for members of about 60,000 households. Related data are also collected about pre-schooling and the general adult population. In addition, NCES funds additional items on education-related topics, such as language proficiency, disabilities, computer use and access, student mobility, and private school tuition.

Beginning in 1972, NCES has conducted longitudinal surveys of students of specified grades who are followed over time.[18] The first such survey, NLS-72, was of students who were high school seniors in the spring of 1972; they were reinterviewed four times over a 14-year period. The first longitudinal study of younger children began in 1998 with a sample of kindergartners, who were followed through 2007 (ECLS-K), and a sample of newborns in 2001, who were followed through 2007 (ECLS-B). The most recent such surveys include ELS:2002, a sample of high school sophomores in 2002 and seniors in 2004, who were followed through 2012; HSLS-09, a sample of students enrolled in 9th grade in 2009, who were followed through 2016; ECLS-K:2011, a sample of kindergartners in 2011, who were followed through 2016; and MGLS:2017, a sample of students enrolled in 6th grade in 2017, who will be followed through 2020. NCES also conducts longitudinal surveys of postsecondary students. These surveys contain rich content, including not only students' academic progress and achievement, but also their social, emotional, and physical development. They typically also include extensive information on the children's homes, classrooms, and school environments.

The NHES is designed to address a wide range of education-related topics.[19] Administrations of the survey were conducted by telephone approximately every 2 years from 1991 through 2007. Because of falling response rates, it was redesigned as a mail survey and administered in 2012 and 2016. Topics covered relevant to educational equity indicators for K–12 have included early childhood program participation (1991, 1995, 2001,

---

[17]See https://www2.census.gov/programs-surveys/cps/techdocs/cpsmar18.pdf and https://www2.census.gov/programs-surveys/cps/techdocs/cpsoct17.pdf; https://nces.ed.gov/surveys/cps.

[18]See https://nces.ed.gov/surveys/hsls09/; https://nces.ed.gov/ecls/; https://nces.ed.gov/training/datauser/COMO_07/assets/COMO_07_Slides.pdf.

[19]See https://nces.ed.gov/nhes.

2005, 2012, 2016); school readiness (1993, 1999, 2007); parent and family involvement in education (1996, 1999, 2003, 2007, 2012, 2016); community service and civic involvement of students in grades 6–12 (1996, 1999); plans for post-high-school education (1999); nonparental care and before- and after-school educational activities (1999, 2001, 2005 [after-school activities only]); and school safety and discipline (1993). Sample sizes have ranged from 7,000 to 21,000 students or parents, depending on the topic.

The NTPS is a biannual sample survey of public K–12 schools, including public charter schools, designed to produce national estimates of teacher, principal, and school characteristics (sample size of about 8,300 schools).[20] Each school in the sample provides a teacher listing form and fills out a school questionnaire, while its principal fills out his or her own questionnaire, and a sample of teachers fills out a teacher questionnaire. The NTPS is a redesign of the Schools and Staffing Survey, which NCES conducted from 1987 to 2011. The NTPS collects data on core topics, including teacher and principal preparation, classes taught, school characteristics, and demographics of the teacher and principal labor force. In addition, each administration of NTPS contains rotating modules on important education topics, such as professional development, working conditions, and evaluation.

---

[20]See http://nces.ed.gov/surveys/ntps.

Appendix B

Assessment of Relevant Publications

In addition to reviewing databases, the committee conducted a broad review of organizations that compile data and prepare reports related to equity in K–12 education. Our review considered both government agencies and nongovernmental organizations; we also looked at efforts that targeted equity from the outset as their raison d'etre (such as briefs from the Civil Rights Data Collection [CRDC] and *Race for Results*—see below), in addition to efforts that included relevant indicators without a specific equity focus. The review identified 19 organizations that publish relevant reports: see Table B-1.

The reports and briefs of these organizations are intended for a wide spectrum of audiences. Some of the organizations are involved in all the steps of producing indicator reports, from collecting data to reporting the results (e.g., NCES). Others make use of data collected by government agencies to develop their own indicators and associated reports (e.g., Child Trends). Still others make use of indicators developed by others to include in their own reports. Some organizations publish reports on a regular basis, most often annually (e.g., *Kids Count*); others publish briefs when the findings warrant (e.g., Child Trends, CRDC).

In addition to the list of organizations in Table B-1, the Pell Institute for Higher Education has a publication series on indicators of equity in higher education, which we do not include because the committee's charge is for K–12 education. The Pell Institute series covers such topics as high school students with college potential, who enrolls in college and what types of college, whether financial aid levels the playing field, how students pay for college, and graduation rates and early income outcomes. Indicators are

TABLE B-1 Organizations and Their Reports with Indicators of Educational Equity

Organization	Report
Alliance for Excellence in Education	Graduation rates
American Youth Policy Forum	Various briefs
Annie E. Casey Foundation	Various briefs and reports, including: *Kids Count Data Book* and *Race for Results*
Child Trends	Various reports and briefs
Council of Chief State School Officers	Various reports and briefs
Council of the Great City Schools	Various reports and briefs, including: *Academic Key Performance Indicators: 2018 Report*
Council of State Governments, Public Policy Research Groups	Various reports and briefs
Education Trust	Various reports and briefs
Education Law Center and Rutgers University	*Is School Funding Fair: A National Report Card on Funding Fairness*
Federal Interagency Forum on Child and Family Statistics	*America's Children: Key National Indicators of Well-Being* and *America's Children in Brief: Key National Indicators of Well-Being*
Princeton University-Brookings Institution Collaboration	*The Future of Children*: Policy Briefs and Special Topic Volumes, published spring and fall
Georgetown University, Center on Education and the Workforce	Various reports and briefs
National Assessment Governing Board (for the National Assessment of Educational Progress [NAEP])	*The Nation's Report Card*: Reports of achievement from main NAEP for 4th, 8th, and 12th graders in math, reading, science, and other subjects; Reports of achievement for the Trial Urban District Assessment
National Association of School Boards (NASBE)	Various reports and briefs

TABLE B-1 Continued

Organization	Report
National Center for Education Statistics (NCES)	Various statistical reports and briefs, including: *Condition of Education (Highlights)* and *Status and Trends in the Education of Racial and Ethnic Groups*
National Governors Association	Various reports and briefs
National Institute for Early Education Research (NIEER), Rutgers University	*State of Preschool* yearbooks
U.S. Census Bureau	Various statistical reports and briefs
U.S. Department of Education, Office of Civil Rights	*Civil Rights Data Collection (CRDC) Issue Briefs* (national highlights on varied topics); *CRDC Special Reports,* highlights on English learners, discipline, and educational equity for school districts and schools

presented for student groups identified by race and ethnicity and family income levels, and the Institute's 2018 report provides indicators for states.[1] The European Union has also undertaken work over the past decade on measuring educational equity among member countries for student groups defined by family income and immigrant status.

The committee's review identified reports from seven organizations as meriting in-depth exploration, using the criteria in Box B-1:

- Annie E. Casey Foundation: *Kids Count Data Book* and *Race for Results*
- Council of the Great City Schools: *Academic Key Performance Indicators: 2018 Report*
- Education Law Center and Rutgers University: *Is School Funding Fair? A National Report Card on Funding Fairness*
- Federal Interagency Forum on Child and Family Statistics: *America's Children: Key National Indicators of Well-Being*
- National Center for Education Statistics: *Condition of Education* and *Status and Trends in the Education of Racial and Ethnic Groups*
- National Institute for Early Education Research, Rutgers University: *State of Preschool* yearbooks
- U.S. Department of Education, Office of Civil Rights: Civil Rights Data Collection *First Look Issue Briefs* and *Special Reports*

---

[1]See http://pellinstitute.org/indicators/reports_2018_data.shtml.

BOX B-1
Criteria for Assessing Publications for
Educational Equity Indicators

The first seven criteria are the same as those for our review of data systems in Appendix A; the last three are unique to this list.

1. Published on a regular, frequent basis—at least annually.
2. Available for subnational geographic areas, including states, school districts, and, ideally, schools or school attendance areas, as appropriate.
3. High-quality when assessed on measures of nonsampling error (e.g., accurate reporting of student enrollment) and on measures of sampling error (for survey-based data).
4. Available for groups of children of interest for education equity (see Chapter 2 text), as defined by race and ethnicity, gender, family income (or equivalent measure of socioeconomic resources), disability status, immigrant status, and English language capability.
   a. For immigrant children, indicative of time of entry into the United States to appropriately include/exclude them in equity indicators (e.g., exclude from a high school graduation measure if they arrived only a year before graduation).
   b. For English-language learners, when possible, indicative of the number of years spent in an English-learner program, whether a student waived out of English-learner instruction, and time and type of reclassification to English-proficient status.
5. Measures contextual factors, such as neighborhood income and family type composition for student groups of interest (see Chapter 3).[a]
6. Measures students' educational outcomes for student groups of interest in three domains comprising seven indicators, each with one or more constructs to be measured (see Chapter 4):
   Domain A: Kindergarten readiness
      Indicator 1: Disparities in academic readiness (reading/literacy, numeracy/math skills)
      Indicator 2: Disparities in self-regulation and attention skills
   Domain B: K–12 learning and engagement (measured at multiple levels/grades)
      Indicator 3: Disparities in engagement in schooling (attendance/absenteeism, academic engagement)
      Indicator 4: Disparities in performance in coursework (success in classes, accumulating credits to be on track to graduate, grades/GPA)
      Indicator 5: Disparities in performance on tests (achievement in reading/math/science, learning growth in reading/math/science achievement)
   Domain C: Educational attainment
      Indicator 6: Disparities in on-time high school graduation
      Indicator 7: Disparities in postsecondary readiness (enrollment in college, entry into the workforce, enlistment in the military)

7. Measures school-provided opportunities to learn for student groups of interest in four domains comprising nine indicators, each with one or more constructs to be measured (see Chapter 5):

   Domain D: Extent of racial, ethnic, and economic segregation

       Indicator 8: Disparities in students' exposure to racial, ethnic, and economic segregation (concentrated poverty in schools, racial segregation within and across schools)

   Domain E: Equitable access to high-quality early learning programs

       Indicator 9: Disparities in access to and participation in high-quality pre-K programs (availability and participation in licensed pre-K programs)

   Domain F: Equitable access to high-quality instruction and curricula

       Indicator 10: Disparities in access to effective teaching (teachers' years of experience, teachers' credentials/certification, racial/ethnic diversity of the teaching force)

       Indicator 11: Disparities in access to and enrollment in rigorous coursework (availability/enrollment in advanced, rigorous coursework, availability/ enrollment in Advanced Placement, International Baccalaureate, and dual enrollment programs, availability/enrollment in gifted and talented programs)

       Indicator 12: Disparities in curricular breadth (availability/enrollment in coursework in the arts, social sciences, sciences, and technology)

       Indicator 13: Disparities in access to high quality academic interventions and supports (access to and participation in formalized systems of tutoring or other types of academic supports; access to and participation in appropriate academic content for English-language learners and special education children)

   Domain G: Equitable access to supportive school and classroom environments

       Indicator 14: Disparities school climates (perceptions of safety, academic support, academically-focused culture, teacher-student trust);

       Indicator 15: Disparities in nonexclusionary discipline practices (out-of-school suspensions/expulsion);

       Indicator 16: Disparities in non-academic supports for student success (supports for emotional, behavioral, mental, physical health)

8. Constructed in a manner that is intelligible to users of varying levels of analytic sophistication.

9. Constructed so that it is difficult to "game" the indicator to make a school district or school appear to be more equitable than it is.

10. Feasible to produce on a timely basis (i.e., soon after the underlying data are available).

---

[a]Although we do not propose indicators of context, they would be critical to inform efforts of school systems to work with other sectors to combat root causes of poverty and other factors that adversely affect students' educational attainment.

These reports collectively include indicators that measure students' academic achievement and progress; school systems' mechanisms for monitoring access to the resources all students need; the inputs students bring to school with them from their homes, families, and neighborhoods; and outcomes that extend from pre-K through K–12 and beyond. All of them publish some or all indicators for one or more student groups of interest. However, none of them includes a full set of the K–12 educational equity indicators the committee identifies or the full spectrum of student groups or geographic units of interest. Indeed, most of them are limited in scope, student group coverage, and geographic detail, and almost none of them articulates an equity-focused framework for the selection of indicators to publish. Nonetheless, they illustrate how the topic of K–12 educational equity is currently addressed, identifying work that a set of K–12 educational equity indicators could draw on as well as gaps and data and methodological shortcomings (see Appendix C).

ANNIE E. CASEY FOUNDATION

The Annie E. Casey Foundation's KIDS COUNT Data Center "seeks to enrich local, state and national discussions concerning ways to secure better futures for all children—and to raise the visibility of children's issues through a nonpartisan, evidence-based lens." The center draws on national, state, and local data sources and publications. Annually, the foundation publishes the *KIDS Count Data Book: State Trends in Child Well-Being*, based on the KIDS Count index, which ranks states on 16 indicators in four domains (with four indicators in each domain): (1) economic well-being, (2) education, (3) health, and (4) family and community.

Table B-2 details the four education indicators in the index, showing their source, periodicity, student groups covered, geographic areas covered, and relevance to the committee's proposed indicators.[2] The economic well-being and family and community indicators provide context.

The Annie E. Casey Foundation began publication of its KIDS Count index in 1990 and revised the index in 2012 to include the four domains and 16 indicators. The index is not published for groups of children, except at the national level by race and ethnicity (with data for the states available in the KIDS Count Data Center).

The foundation subsequently developed an index, published in *Race for Results: Building a Path to Opportunity for All Children*, to focus explicitly on children's prospects for success as an adult in each state. The new index, first published in 2014 and updated in 2017, was based on an aspirational

---

[2]The Data Center provides additional education indicators not included in the index: see https://datacenter.kidscount.org/.

goal: all children should grow up in economically successful families; live in supportive communities; and meet developmental, health, and educational milestones. The index comprises 12 indicators that were comparably and regularly collected in every state through surveys sufficient in size to allow valid estimates for the five largest racial groups (African-American, American Indian, Asian American and Pacific Islander, Latino, and White); 4 of the 12 indicators are the 4 education indicators that are part of the KIDS Count index.[3]

---

[3]For the full set of indicators, see https://www.aecf.org/m/resourcedoc/AECF-RaceforResults-2014.pdf#page=33.

TABLE B-2 Education Indicators in the KIDS Count Index

Indicator	Source	Periodicity and Latest Year Available (in parentheses)[a]	Student Groups of Interest for Which Data Are Published (or are Available in the Data Center)	Geographic Areas for Which Data Are Published (or Are Available in the Data Center)	Relevance to Committee's Indicators
Young Children Not in School: percent 3- and 4-year-olds not enrolled in nursery school, pre-K, or K, in prior 3 months	American Community Survey, averaged over 3 years	Annual, 2007-2009 through 2014-2016; will be updated	Race/ethnicity (only published for the nation; available for states)	Nation, states	Indicator 9
4th Grade Reading Achievement Levels: percent 4th graders not proficient in reading	Main NAEP	Every 2 years (2017)	4th graders: race/ethnicity (only published for the nation; available for states)	Nation, states	Indicator 5 for 4th grade
8th Grade Math Achievement Levels: percent 8th graders not proficient in math	Main NAEP	Every 2 years (2017)	8th graders: race/ethnicity (only published for the nation; available for states)	Nation, states	Indicator 5 for 8th grade
High School Students Not Graduating on Time: (ACGR)[b]	Common Core of Data	Each school year from 2010-2011 (different rate used earlier) (2015-2016)	Race/ethnicity (only published for the nation; available for states)	Nation, states (available for school districts and schools)	Indicator 8

[a]Latest year at time of latest publication (see SOURCE below).
[b]For the definition of ACGR (Adjusted Cohort Graduation Rate), see https://nces.ed.gov/blogs/nces/post/what-is-the-difference-between-the-acgr-and-the-afgrfor definition.
NOTE: NAEP, National Assessment of Educational Progress.
SOURCE: Information from The Annie E. Casey Foundation (2018).

COUNCIL OF THE GREAT CITY SCHOOLS

The Council of the Great City Schools began a project in 2014 to develop key academic and cost performance indicators for 74 urban districts that are members of the Council. The first phase was a pilot study in which 8 member districts initially participated, followed by a second phase in which general education, special education, English-language learner, finance, and research department representatives narrowed down a preliminary set of 200 indicators to about 58 indicators. In a third phase, data were collected and results published for about a dozen academic indicators for about 50 member districts for the 2015-2016 school year.[4] The data have been updated for the 2016-2017 school year, again for about 50 districts (although all 74 were asked to participate).

Table B-3 describes the academic key performance indicators in the latest Council report for 2016-2017, showing their source, periodicity, student groups covered, geographic areas covered, and relevance to the committee's proposed indicators. The data are still considered preliminary and so are not immediately relevant to the committee's set of education equity indicators. However, the processes and steps the Council followed to assess the feasibility and burden of providing the indicators and to reach consensus among the member districts as to their value may carry lessons for implementation of the committee's recommended set.

---

[4] *Academic Key Performance Indicators: Pilot Report* (October 2017); see https://www. cgcs.org/site/default.aspx?PageType=3&ModuleInstanceID=1044&ViewID=C9E0416E-F0E7-4626-AA7B-C14D59F72F85&RenderLoc=0&FlexDataID=3545&PageID=836. Work on cost indicators was deferred until after the academic indicators were fully developed.

TABLE B-3 Indicators in *Academic Key Performance Indicators: 2018 Report*

Indicator	Source	Periodicity and Latest Year Available (in parentheses)[a]	Student Groups of Interest for Which Data Are Published	Geographic Areas for Which Data Are Published	Relevance to Committee's Indicators
Pre-K Enrollment as a Percentage of Kindergarten Enrollment	Council of the Great City Schools online survey and Excel spreadsheets	Annual (potentially) (2016-2017 school year)	Total, Black males, Hispanic males, free or reduced-price school lunch eligible students, students with disabilities,[b] English-language learners	74 urban school districts that are Council members (potentially—current participation about 50 districts)	Indicator 9
Percent of 4th Graders Proficient in Math and Reading and Below Basic	NAEP	Every 2 years (2017)	Total, free or reduced-price lunch students, students with disabilities,[b] race/ethnicity (black, Hispanic, white) by gender and by free or reduced-price school lunch eligibility	27 Council members (as of 2017) that participate in NAEP Trial Urban District Assessment	Indicator 5
Percent of 8th Graders Proficient in Math and Reading and Below Basic	NAEP	Every 2 years (2017)	See Percent of 4th Graders above	See Percent of 4th Graders above	Indicator 5
Percent of 9th Graders Completing Algebra I by End of 9th Grade, by Percent in Grades 7, 8, 9	See Pre-K Enrollment above	See Pre-K Enrollment above	See Pre-K Enrollment above	See Pre-K Enrollment above	Indicators 4 and 11

continued

TABLE B-3 Continued

Indicator	Source	Periodicity and Latest Year Available (in parentheses)[a]	Student Groups of Interest for Which Data Are Published	Geographic Areas for Which Data Are Published	Relevance to Committee's Indicators
Percent of 9th Graders Failing One or More Courses	See Pre-K Enrollment above	See Pre-K Enrollment above	See Pre-K Enrollment above	See Pre-K Enrollment above	Indicator 4
Percent of 9th Graders with B Average GPA or Better	See Pre-K Enrollment above	See Pre-K Enrollment above	See Pre-K Enrollment above	See Pre-K Enrollment above	Indicator 4
Percent of 3rd, 6th, 8th, and 9th Graders Who Missed School by Days Absent (5-9, 10-19, 20+)	See Pre-K Enrollment above	See Pre-K Enrollment above	See Pre-K Enrollment above	See Pre-K Enrollment above	Indicator 3
Percent of Students with Out-of-School Suspensions by Number of Days Suspended (1-5, 6-10, 11-19, 20+)	See Pre-K Enrollment above	See Pre-K Enrollment above	See Pre-K Enrollment above	See Pre-K Enrollment above	Indicator 15
Instructional Days Missed per Student Due to Suspensions	See Pre-K Enrollment above	See Pre-K Enrollment above	See Pre-K Enrollment above	See Pre-K Enrollment above	Indicator 15
Percent of Secondary Students Who Took One or More AP Courses[c]	See Pre-K Enrollment above	See Pre-K Enrollment above	See Pre-K Enrollment above	See Pre-K Enrollment above	Indicator 11
Percent of AP Scores that Passed (scores of 3 or higher)		See Pre-K Enrollment above			Indicator 4

4-Year Cohort High School Graduation Rate	See Pre-K Enrollment above	See Pre-K Enrollment above	See Pre-K Enrollment above	See Pre-K Enrollment above	See Pre-K Enrollment above	Indicator 6

[a]To date, the Council of the Great City Schools has published reports for 2016 and 2018; whether the reports will be annual, biannual, or on issued on another schedule is not known. Except for NAEP scores, all of the indicators are potentially available on an annual basis.

[b]Disability status not defined.

[c]AP-equivalent participation and participation in college courses also collected but not published.

NOTES: AP, Advanced Placement; GPA, grade point average; NAEP, National Assessment of Educational Progress.

SOURCE: Information from *Academic Key Performance Indicators 2018 Report* (October 2018), available: https://www.cgcs.org/site/handlers/filedownload.ashx?moduleinstanceid=1044&dataid=3545&FileName=Final%20KPI%20Report%202017.pdf.

EDUCATION LAW CENTER AND RUTGERS UNIVERSITY

The Education Law Center and Rutgers University have collaborated since 2010 on an annual publication that assesses the fairness of funding for K–12 public education state by state. *Is School Funding Fair? A National Report Card on Funding Fairness* presents data on four indicators of "fairness" and three indicators of resource allocation. Box B-2 presents the principles on which the published fairness and resource allocation indi-

---

BOX B-2
Principles Used to Derive Indicators in
Is School Funding Fair?

- Varying levels of funding are required to provide equal educational opportunities to children with different needs.
- The costs of education vary based on geographic location, regional differences in teacher salaries, school district size, population density, and various student characteristics. It is critical to account for as many of these variables as possible, given the availability of reliable data.
- The level of funding should increase relative to the level of concentrated student poverty—that is, state finance systems should provide more funding to districts serving larger shares of students in poverty . . . .
- Student poverty—especially concentrated student poverty—is the most critical variable affecting funding levels. Student and school poverty correlates with, and is a proxy for, a multitude of factors that increase the costs of providing equal educational opportunity—most notably, gaps in educational achievement, school district racial composition, English-language proficiency, homelessness, and student mobility. State finance systems should deliver greater levels of funding to higher-poverty settings, while controlling for differences in other cost factors.
- While the distribution of funding to account for student need is crucial, the overall funding level in states is also a significant element in fair school funding. Without sufficient base or foundational funding, even a progressively funded system will be unable to provide equitable educational opportunities.
- The sufficiency of the overall level of funding in any state can be assessed based on comparisons to other states with similar conditions and similar characteristics. Using available national data, average differences in state and local revenues between states, as well as within states, can be projected and indexed to compare expected state and local revenues per pupil under a given set of conditions. These expected values are derived from a statistical model that predicts funding levels while controlling for various school district characteristics. These predicted funding levels allow for more direct comparison of districts having similar characteristics across states.

SOURCE: Baker, Farrie, and Sciarra (2018, p. 3). Reprinted with permission from the Education Law Center.

cators are based; Table B-4 describes each of the seven indicators, showing its source, periodicity, student groups covered, geographic areas covered, and relevance to the committee's proposed indicators.

TABLE B-4 Indicators in *Is School Funding Fair? A National Report Card on Funding Fairness*

Indicator	Source	Periodicity	Student Groups of Interest for Which Data Are Published	Geographic Areas for Which Data Are Published	Relevance to Committee's Indicators
Fairness Measure #1, Funding Level—adjusted per pupil funding[a]	Census Bureau Annual Survey of School System Finances; SAIPE; Taylor's extended NCES Comparable Wage Index[b] CCD; Census Bureau population estimates	Annual, beginning 2010; publication lags by 2-3 years (e.g., 2018 edition presents 2015 estimates)	N.A., but indicator takes account of student poverty	States (states are ranked)	Indicator 7.3
Fairness Measure #2, Funding Distribution—progressive, regressive, or flat distribution[c]	See fairness measure #1	See Fairness Measure #1	N.A., but indicator takes account of school district poverty	States (states are grouped into six grades, A-F)	Indicators 13, 16
Fairness Measure #3, Fiscal Effort—local and state education spending relative to (a) gross state product (GSP) and (b) state aggregate personal income (SPI)	BEA GSP and SPI series; Urban Institute-Brookings Institution Tax Policy Center Data Query System	See Fairness Measure #1	N.A.	States (states are grouped into six grades, A-F)	N.A.

continued

Fairness Measure #4, Coverage—proportion of school-aged children (6-16) attending the state's public schools averaged with ratio of household median income of public school students to other students	ACS (3-year public-use microdata samples)	See Fairness Measure #1	N.A.	States (states are ranked)	N.A.
Resource Allocation Measure #1, Early Childhood Education—enrollment of 3- and 4-year-olds in early childhood education programs by income level	ACS (3-year public-use microdata samples)	See Fairness Measure #1	N.A, but gives extra weight to enrollment of children in low-income families	States (states are ranked)	Indicator 9
Resource Allocation Measure #2, Wage Competitiveness—ratio of teacher to nonteacher wages	ACS (3-year public-use microdata samples)	See Fairness Measure #1	N.A.	States (states are ranked)	N.A.

TABLE B-4 Continued

Indicator	Source	Periodicity	Student Groups of Interest for Which Data Are Published	Geographic Areas for Which Data Are Published	Relevance to Committee's Indicators
Resource Allocation Measure #3, Teacher-to-Student Ratios—fair: higher ratios in high poverty districts; unfair: comparable or lower ratios in, high poverty districts	CCD—Local Education Agency Universe Survey	See Fairness Measure #1	N.A., but takes account of school district poverty	States (states are ranked)	Indicator 6.1

[a]The adjustment is based on a model that predicts average funding levels (state plus local), controlling for student poverty, regional wage variation, and school district size and density.

[b]The NCES Comparable Wage Index (CWI) is a measure of regional variations in the salaries of college graduates who are not educators; see updated NCES CWI data beginning in 2005 based on 3-year average ACS data at http://bush.tamu.edu/research/faculty/Taylor_CWI.

[c]Progressive: high-poverty (30%+) districts receive in 2005 based on 3-year average ACS data receive at least 5 percent additional funds over low-poverty (0%) districts; regressive: high-poverty districts receive 5 percent less funding than low-poverty districts; flat: in between.

NOTES: ACS, American Community Survey (Census Bureau program); BEA, Bureau of Economic Analysis; CCD, Common Core of Data (NCES program); N.A., not applicable; SAIPE, Small Area Income and Poverty Estimates (Census Bureau program).

SOURCE: Information from Baker, Farrie, and Sciarra (2018).

FEDERAL INTERAGENCY FORUM ON
CHILD AND FAMILY STATISTICS

The Federal Interagency Forum on Child and Family Statistics fosters coordination and collaboration among 23 federal agencies that produce or use statistical data on children and families. Through the *America's Children* series, the forum makes federal data on children and families available in a nontechnical, easy-to-use format in order to stimulate discussion among data providers, policy makers, and the public.

The forum publishes an annual report on the well-being of children and families, alternating between the longer *America's Children: Key National Indicators of Well Being* and *America's Children in Brief.* The full reports and the forum's website provide statistics on 41 indicators:[5]

> . . . that must meet the following criteria: easy to understand by broad audiences; objectively based on reliable data with substantive research connecting them to child well-being; balanced, so that no single area of children's lives dominates the report; measured regularly, so that they can be updated and show trends over time; and representative of large segments of the population, rather than one particular group.

The indicators cover seven domains: family and social environment, economic circumstances, health care, physical environment and safety, behavior, education, and health. Because the briefer reports do not cover every domain every year, any one domain, such as education, is published at least biannually but not necessarily annually. Moreover, not all indicators within a domain are available annually.

Of the 41 indicators, 6 are devoted to education. Table B-5 defines each of the six, indicating source, periodicity, student groups covered, geographic areas covered, and relevance to the committee's indicators. Indicators in the family and social environment and economic circumstances domains provide context.

---

[5]Forum on Child and Family Statistics; available: https://www.childstats.gov/americaschildren/index.asp.

TABLE B-5 Education Indicators in *America's Children*

Indicator	Source	Periodicity and Latest Year Available (in parentheses)[a]	Student Groups of Interest for Which Data Are Published (or Can Be Made Available)	Geographic Areas for Which Data Are Published (or Can Be Made Available)	Relevance to Committee's Indicators
1—Family Reading to Young Children: percent children ages 3–5 read to three or more times in last week by a family member	NHES	1993, 1995, 1996, 1999, 2001, 2005, 2007, 2012, 2016—future periodicity unknown	Gender, race/ ethnicity, poverty status, family type, mother's highest level of education, and mother's employment status	Nation, 4 regions	N.A.
2—Math and Reading Achievement: average math and reading scale scores of 4th, 8th, and 12th graders	Main NAEP	Every 2 years (2015)	4th graders: by gender and by race/ ethnicity; 8th and 12th graders: by gender, by race/ethnicity, and by parents' education	Nation (states and some large cities available)	Indicator 5
3—High School Academic Course-Taking: percent public high school students enrolled in selected mathematics and science courses[b]	CCD; CRDC	Every other school year (2013–2014)	Type of course: by gender, by race/ ethnicity, and by gender by race/ ethnicity	Nation (states, school districts, schools available)	Indicator 11

Indicator	Data Source	Frequency (Latest Year)	Breakdown	Geographic Level	Reference
4—High School Completion: percent adults ages 18-24 who completed high school (including a GED)	CPS School Enrollment Supplement	Annual (2015)	Race/ethnicity	Nation (regions are available, as are states with 3-year averaging)	Indicator 6
5—Youth by School Enrollment and Work Status: percent youth ages 16-19 (school includes high school and college)	CPS (monthly for school months)	Annual (2016)	Age (16-17, 18-19): by gender, by race/ethnicity, and by enrollment and working status	Nation (regions are available, as are states with 3-year averaging)	Indicator 7
6—College Enrollment: percent high school completers enrolled in college the following fall	CPS School Enrollment Supplement	Annual (2015)	Gender, race/ethnicity, and income level (low, middle, high)	Nation (regions are available, as are states with 3-year averaging)	Indicator 7

[a]Latest year at time of latest publication (see SOURCE below).

[b]Algebra 1, geometry, algebra 2, advanced mathematics, calculus, advanced placement (AP) math, biology, chemistry, physics, AP science.

NOTES: CCD, Common Core of Data (NCES program); CPS, Current Population Survey (Census Bureau program); CRDC, Civil Rights Data Collection (Office of Civil Rights program); GED, general education diploma; N.A., not applicable; NAEP, National Assessment of Educational Progress; NHES, National Household Education Survey (NCES program).

SOURCE: Information from Federal Interagency Forum on Child and Family Statistics (2017).

NATIONAL CENTER FOR EDUCATION STATISTICS

Condition of Education

Condition of Education reports are issued annually by NCES in compliance with a congressional mandate. The reports contain indicators on the state of education in the United States, from pre-kindergarten through postsecondary education, as well as labor force outcomes and international comparisons. For pre-K through grade 12, there are indicators of family characteristics, enrollment, teachers and staff, assessments, high school completion, and school finance. More detailed information, on which the reports are based, is available in the annual *Digest of Education Statistics*.[6]

The data for these indicators are obtained from many different providers—including students and teachers, state education agencies, local elementary and secondary schools, and colleges and universities—using surveys and compilations of administrative records. Most indicators in the reports summarize data collected by surveys conducted by NCES or by the Census Bureau with support from NCES, such as the American Community Survey (ACS) and the Current Population Survey (CPS).

The *Condition of Education* includes an *At a Glance* section, which allows readers to quickly make comparisons within and across indicators, and a *Highlights* section, which "spotlights" key findings for a few of the indicators. Table B-6 defines topic areas for which regularly collected indicators are provided for pre-K through grade 12 in the latest (2018) *Condition of Education*, indicating source, periodicity, student groups covered, geographic areas covered, and relevance to the committee's indicators.[7]

---

[6]See https://nces.ed.gov/programs/digest/.

[7]In addition to the indicators shown, *Condition of Education* reports on children's access to and use of the Internet, measured in CPS supplements in October 2010 and July 2015; family involvement in educational activities outside school, measured in 2012 and 2016 in NHES; school crime and safety, measured periodically in the NCES Survey on School Crime and Safety; public school teacher turnover, measured in a 2012-2013 Teacher Follow-up Survey to the 2011-2012 Schools and Staffing Survey (SASS); characteristics of public school principals, measured periodically in SASS (latest estimates for 2011-2012); public school principal turnover, measured in a 2012-2013 Principal Follow-up Survey to the 2011-2012 SASS; trends in reading and math scale scores for 9-, 13-, and 17-year-olds from 1971 through 2012, measured in long-term trend NAEP; technology and engineering literacy for 8th graders, measured by NAEP in 2014; and high school graduates by completion of math and science courses, measured in 2000 and 2009 NAEP high school transcript studies.

TABLE B-6 Indicators in the *Condition of Education*

Indicator	Source	Periodicity and Latest Year Available (in parentheses)[a]	Student Groups for Which Data Are Published[b]	Geographic Areas for Which Data Are Published[c]	Relevance to Committee's Indicators
Characteristics of Children's (under age 18) Families—various percentages of children	ACS	Annual (2016)	Race/ethnicity: by parents' educational attainment, by family type, and by poverty status; living in poverty by race/ethnicity: by family type, and by parents' educational attainment	Nation; states for percentage living in poverty	Contextual factors
Pre-K and K Enrollment—percent ages 3, 4, 5 enrolled	CPS SES	Annual (2017)	3- to 5-year-old enrollment by full- or part-day: by race/ethnicity, and by parents' educational attainment	Nation	Indicator 9
Elementary and Secondary Enrollment—percent enrolled (any type of school); number enrolled and projected in public school	CPS SES for percent enrolled; CCD for number enrolled	Annual (2016, projections through 2027)	Percent enrolled: by age (3-4, 5-6, 7-13, 14-15, 16-17, 18-19); Enrolled and projected: by school level (elementary, secondary)	Nation; states by actual and projected percentage change in public K–12 enrollment	Denominator for various indicators

continued

TABLE B-6 Continued

Indicator	Source	Periodicity and Latest Year Available (in parentheses)[a]	Student Groups for Which Data Are Published[b]	Geographic Areas for Which Data Are Published[c]	Relevance to Committee's Indicators
Public Charter School Enrollment—number by school level; various percentages	CCD	Annual (2015-2016)	Percent public charter school distribution by size; percent student distribution by race/ethnicity	Nation; states by percent public students enrolled in charters	N.A.
Private School Enrollment—percent of all K–12 students in private schools; various numbers and percentages	NCES Private School Universe Survey	Every 2 years (2015-2016)	Number enrolled by grade level (pre-K-8, 9-12) and by orientation (Catholic, other religious, nonsectarian); percent student distribution: by school level by orientation, and by race/ethnicity by orientation	Nation	N.A.

| English-Language Learners (ELL) in Public Schools—various percentages | CCD, CRDC | Every 2 years (fall 2015) | Percent ELL of total pre-K through grade 12: by grade, and by school locale (city, suburban, town, rural); number and percent distribution of ELL students by home language | Nation; states by percent of public school enrollment | Indicator 7.3 |
| Children and Youth (Ages 3-21) with Disabilities—percent receiving special education services under IDEA, Part B | Office of Special Education Programs, IDEA database | Annual (2015-2016) | Percent served: ages 3-21 by type (10 types), and by race/ethnicity; ages 6-21 by time in general classes; ages 14-21 exiting school by reason (regular diploma, alternative certificate) by race/ethnicity | Nation | Indicator 13 |

continued

TABLE B-6 Continued

Indicator	Source	Periodicity and Latest Year Available (in parentheses)[a]	Student Groups for Which Data Are Published[b]	Geographic Areas for Which Data Are Published[c]	Relevance to Committee's Indicators
Characteristics of Traditional and Charter Public Schools—various percentages	CCD	Annual (2015-2016)	Percent traditional and public charter schools: by level, by race/ethnicity concentration (50%+ white, black, Hispanic), by eligible for free or reduced-price lunch (0-25%, 25.1-50%, 50.1-75%, 75%+), and by school locale	Nation	Indicator 8
Concentration of Students Eligible for Free or Reduced-Price Lunch—various percentages	CCD	Annual (2015-2016)	Percent students by "poverty" (categories defined by quartiles of school lunch eligibility): by race/ethnicity, and by school locale	Nation	Indicator 8

Characteristics of School Teachers—various percentages	NTPS	Every 2 years (2015-2016)	Percent public school teachers: by gender by level, by race/ethnicity, by level by college degree/teaching certificate, and by years of teaching experience by average base salary and by highest degree	Nation	Indicator 10
Reading Performance—average scale scores/four achievement levels for 4th, 8th, and 12th graders	Main NAEP	4th, 8th graders: every 2 years (2017); 12th graders: periodically (2015)	Scale scores: by gender, by ELL status, by race/ethnicity, and by school poverty (based on quartiles of students eligible for free or reduced-price lunch)	Nation; change in 4th and 8th grade scale scores by state	Indicator 5
Mathematics Performance—see Reading Performance, above	Main NAEP	See Reading Performance, above	See Reading Performance, above	See Reading Performance, above	Indicator 5
Science Performance—average scale scores for 4th, 8th, and 12th graders	Main NAEP	Periodically (2015)	Gender and race/ethnicity	Nation; change in scale scores by state	Indicator 5

continued

TABLE B-6 Continued

Indicator	Source	Periodicity and Latest Year Available (in parentheses)[a]	Student Groups for Which Data Are Published[b]	Geographic Areas for Which Data Are Published[c]	Relevance to Committee's Indicators
High School Graduation Rates—adjusted cohort graduation rates	Consolidated State Performance Report (in EDFacts)	Annual (2015-2016)	Race/ethnicity	Nation; states	Indicator 6
Status Dropout Rates— percent 16- to 24-year- olds not enrolled in school and lacking a diploma or GED	CPS SES	Annual (2016)	Gender, years of school completed, and race/ethnicity by native/foreign born	Nation	Indicator 7
Revenue Sources—percent federal, state, local; state revenue and property tax revenue as percent of total	CCD	Annual (2014-2015)	N.A.	Nation; states for all but revenue source	Indicators 13, 16
Expenditures—current expenditures, interest, and capital outlays per student; percent of current expenditures for salaries, benefits, purchased services, supplies	CCD	Annual (2014-2015)	N.A.	Nation	Indicators 13, 16

[a]Latest year at time of latest publication (see SOURCE below).
[b]Additional student group detail available in the annual *Digest of Education Statistics*.
[c]Additional geographic detail available in the annual *Digest of Education Statistics*.

NOTES: The information covers public schools unless otherwise noted. ACS, American Community Survey; CCD, Common Core of Data; CPS SES, Current Population Survey School Enrollment Supplement; CRDC, Civil Rights Data Collection; ELL, English-language learner; GED, general education diploma; IDEA, Individuals with Disabilities Education Act; N.A., not available; NAEP, National Assessment of Educational Progress; NCES, National Center for Education Statistics: NTPS, National Teacher and Principal Survey.
SOURCE: Information from McFarland et al. (2018).

Status and Trends in the Education of Racial and Ethnic Groups

NCES began issuing reports that focus on the educational progress and challenges facing students in the United States by race and ethnicity in 2003 with *Status and Trends in the Education of Blacks* and *Status and Trends in the Education of Hispanics*. These reports were followed in 2005 by *Status and Trends in the Education of American Indians and Alaska Natives* (updated in 2008). In 2007, 2010, 2015, 2016, 2017, and 2019, NCES published *Status and Trends in the Education of Racial and Ethnic Groups*; presumably, this report will be a continuing series.

Table B-7 defines topic areas for which regularly collected indicators are provided for pre-K through grade 12 in the latest *Status and Trends* report, indicating source, periodicity, student groups covered, geographic areas covered, and relevance to the committee's indicators. Note that some indicators in *Status and Trends* are similar to those in the *Condition of Education*, but *Status and Trends* contains some additional indicators.[8]

---

[8]*Status and Trends in the Education of Racial and Ethnic Groups, 2018* (February 2019) also reports on: childcare arrangements for children under 6 (measured periodically in the National Household Education Survey); high school course taking and whether earned AP or IB credits (measured in the High School Longitudinal Study of 2009); and school safety (measured in the Youth Risk Behavior Survey and the School Crime Supplement to the National Crime Victimization Survey for 2015).

TABLE B-7 Indicators in *Status and Trends in the Education of Racial and Ethnic Groups*

Indicator	Source	Periodicity and Latest Year Available (in parentheses)[a]	Student Groups for Which Data Are Published	Geographic Areas for Which Data Are Published	Relevance to Committee's Indicators
Demographics—percent 5- to 17-year-olds; percent distribution of under 18 by nativity, family type, living in poverty and mother-only households living in poverty (official, supplemental poverty measures)	Census Bureau Population Estimates; ACS; CPS ASEC	Annual (2017, 2016)	Race/ethnicity (detailed Asian, Hispanic groups for nativity, family type, official poverty)	Nation	Contextual factors
Elementary and Secondary Enrollment—percent enrolled in public schools (pre-K to 12); distribution by region, traditional or charter; private school distribution by type of school	CCD; Private School Universe Survey	Annual (2015)	Race/ethnicity	Nation, four regions (public school enrollment)	Denominator for various indicators
English-Language Learners (in public schools)—number; percent of total enrollment	CCD; EDFacts	Annual (2015)	Race/ethnicity	Nation	Indicators 13, 16

continued

TABLE B-7 Continued

Indicator	Source	Periodicity and Latest Year Available (in parentheses)[a]	Student Groups for Which Data Are Published	Geographic Areas for Which Data Are Published	Relevance to Committee's Indicators
Students with Disabilities—percent of students ages 3-21 served under IDEA, Part B; distribution by type of disability; percent ages 14-21 who exited school by reasons	CCD; Office of Special Education Programs, Individuals with Disabilities Education Act (IDEA) database	Annual (2015-2016, 2014-2015)	Race/ethnicity	Nation	Indicators 13, 16
Reading Achievement—average scale score, 4th, 8th, 12th grade	Main NAEP	Every 2 years, 4th, 8th grade (2017); periodically, 12th grade (2015)	Race/ethnicity	Nation	Indicator 5
Mathematics Achievement—see Reading Achievement, above	Main NAEP	See Reading Achievement, above	Race/ethnicity	Nation	Indicator 5
Absenteeism and Achievement—percent 8th graders absent by number of days; average math/reading scale scores by number days absent	Main NAEP	Every 2 years (2017)	Race/ethnicity	Nation	Indicator 3

Retention, Suspension, and Expulsion—percent retained in grade by school level; percent received out-of-school suspensions	CPS SES; CRDC	Annual (retentions, 2016); every 2 years (suspensions, 2013-2014)	Race/ethnicity by gender (suspensions)	Nation	Indicator 15
High School Status Dropout Rates—percent of 16- to 24-year-olds dropping out	ACS	Annual (2016)	Race/ethnicity: by gender and by nativity; Hispanic and Asian by subgroup	Nation	Indicator 7
High School Status Completion Rates— percent of 18- to 24-year-olds completing high school	CPS SES	Annual (2016)	Race/ethnicity and Hispanic/non-Hispanic by recency of immigration	Nation	Indicator 7

[a]Latest year at time of latest publication (see SOURCE below).
NOTES: ACS, American Community Survey; CCD, Common Core of Data; CPS ASEC, Current Population Survey Annual Social and Economic Supplement; CPS SES, Current Population Survey School Enrollment Supplement; CRDC, Civil Rights Data Collection; NAEP: National Assessment of Educational Progress.
SOURCE: Data from de Brey et al. (2019).

NATIONAL INSTITUTE FOR EARLY EDUCATION RESEARCH

NIEER at Rutgers University in 2003 began issuing annual reports on the extent and quality of state pre-K education for children ages 3 and 4. The latest yearbook (Friedman-Krauss et al., 2018, p. 5) notes that, as of the 2001-2002 school year, just two states had pre-K programs that served greater than 50 percent of their 4-year-olds, and 13 states had no state-funded pre-K program. As of 2017, 10 states served more than 50 percent of their 4-year-olds, and only 7 states had no state-funded program.

The NIEER yearbooks provide indicators for access, resources, and quality of pre-K programs for the 50 states and the District of Columbia. States are ranked on measures of access and resources, and the number of quality benchmarks they meet is totaled (from 1 to 10). Table B-8 defines topic areas and indicators for each state in the latest (2017) *State of Preschool* yearbook, indicating source, periodicity, student groups covered, geographic areas covered, and relevance to the committee's indicators. Because pre-K programs vary among and within states, the state profiles in the NIEER yearbooks contain text explaining each state's programs—for example, whether they are offered in all school districts or those meeting a poverty criterion, hours of operation, teacher qualifications, whether they partner with Head Start, and other pertinent information.

continued

TABLE B-8 Indicators in the *State of Preschool* Yearbooks

Indicator	Source	Periodicity and Latest Year Available[a]	Student Groups for Which Data Are Published	Geographic Areas for Which Data Are Published	Relevance to Committee's Indicators
Access (1)— percent 3-year-olds and percent 4-year-olds enrolled in state-funded pre-K, Head Start, Special Education, and other/none	Census Bureau Population Estimates; NIEER State Survey; Department of Education (Special Education); DHHS (Head Start)	Annual (2017)	Special education enrollment (includes unduplicated count of disabled students under IDEA Preschool Grants program)	States	Indicator 9
Access (2)—percent school districts that offer state pre-K program and income requirement	NIEER State Survey	Annual (2017)	N.A.	States	Indicator 9
Access (3)—minimum hours of operation and operating schedule (e.g., school year)	NIEER State Survey	Annual (2017)	N.A.	States	Indicator 9
Quality Standards Checklist—number and which of 10 standards met[b]	NIEER State Survey	Annual (2017)	N.A.	States	Indicator 9

TABLE B-8 Continued

Indicator	Source	Periodicity and Latest Year Available[a]	Student Groups for Which Data Are Published	Geographic Areas for Which Data Are Published	Relevance to Committee's Indicators
Resources—total state spending; state Head Start spending; state spending per child enrolled; all spending per child enrolled[c]	NEA state surveys[d]	Annual (2017)	N.A.	States	Indicator 9

[a]Latest year at time of latest publication (see SOURCE below).

[b]The 10 policy areas and latest benchmarks are: (1) early learning and development standards: comprehensive, aligned, supported, culturally sensitive; (2) curriculum supports: approval process and supports; (3) teacher degree: B.A.; (4) teacher specialized training: specializing in pre-K; (5) assistant teacher degree: Child Development Associate or equivalent credential; (6) staff professional development: at least 15 hours/year, individual development plans, coaching; (7) maximum class size: 20 children or fewer; (8) staff-child ratio: 1:10 or lower; (9) screening and referral: vision, hearing, and health screenings and referral; and (10) monitoring/continuous quality improvement system: structured classroom observation, program improvement plan. Area/benchmark (2) is new as of 2015–2016; a previous area/benchmark related to meals has been discontinued.

[c]Spending includes current operating expenditures plus annual capital outlays and interest on school debt.

[d]In *Rankings and Estimates: Rankings of the States 2016 and Estimates of School Statistics 2017*; see http://www.nea.org/home/73145.htm [April 2019].

NOTES: DHHS, Department of Health and Human Services; IDEA: Individuals with Disabilities Education Act; N.A., not available; NEA, National Education Association; NIEER, National Institute for Early Education Research, Rutgers University.

SOURCE: Information from Friedman-Krauss et al. (2018).

U.S. OFFICE OF CIVIL RIGHTS DATA COLLECTION

The CRDC program in the Office of Civil Rights of the U.S. Department of Education regularly issues "First Look" issue briefs from each biannual cycle of data collection. The topics of these briefs differ from year to year. The two briefs issued to date (in April 2018) from the 2015-2016 data collection are the *STEM* [science, technology, engineering, and math] *Course Taking Issue Brief* and the *School Climate and Safety Issue Brief*. The brief on STEM course taking presents 10 figures: one example is a bar graph of the percentage of high school enrollment by race and ethnicity; the percentage enrolled in algebra 1 for grades 9-10 and 11-12, by race and ethnicity; and the percentage passing algebra 1 for grades 9-10 and 11-12, by race and ethnicity.

The CRDC also provides ready access through a search feature to three special reports for school districts and schools: *English Learner Report, Discipline Report,* and *Educational Equity Report,* which are provided in Excel spreadsheets. Underlying the CRDC issue briefs and special reports are detailed tables of all data elements in the CRDC program for school districts and schools, together with summaries for states and the nation. We do not further describe the particular elements in the special reports or underlying CRDC database because, while very easy to access, they are not presented in the form of regularly published indicators as in the reports described above.[9]

---

[9]For more information about the CRDC and ways to access the data, see https://www2.ed.gov/about/offices/list/ocr/data.html?src=rt.

Appendix C

Data and Methodological Opportunities and Challenges for Developing K–12 Educational Equity Indicators

To make the committee's conclusions and recommendations more concrete, this appendix illustrates for each of the committee's seven recommended domains and 16 indicators the data sources and methods that could be used to measure relevant constructs in appropriate ways. In some instances, the currently available data support proxy measures rather than measures that more directly capture the recommended indicator, or good measures are available but not for all student groups of interest, or the data are very sparse at the scale needed for an educational equity system—that is, comparable, high-quality information nationwide for the nation, states, school districts, and schools. When possible, we indicate ways in which available data could be enhanced or scaled up to fill the gaps. We also note when indicators are included in the publications reviewed in Appendix B.

DOMAIN A: KINDERGARTEN READINESS

The research literature amply supports the importance of kindergarten readiness, both academically and behaviorally, for children's continued educational success (see Chapter 4). Yet this is a domain for which the data are sparse for the committee's two proposed indicators: academic readiness (Indicator 1) and behavioral readiness (Indicator 2).

Indicator 1: Disparities in Academic Readiness

At present, there is no satisfactory data source for developing measures of the two constructs under this indicator—reading/literacy skills and numeracy/math skills—for all levels of needed geography and student groups. Although some states and districts in the country currently assess the literacy and numeracy skills of their entering kindergarten students, there are no broadly used assessments that could provide comparable results nationwide.

The Early Childhood Longitudinal Study: Kindergarten Class of 2010–2011 (ECLS-K:2011; see Appendix A) of the National Center for Education Statistics (NCES) included assessments of sampled students entering kindergarten that school year for reading, math, and science. The assessments contained items developed for the ECLS, items adapted from commercial assessments, and items adapted from other NCES studies within assessment frameworks based on the National Assessment of Educational Progress (NAEP; see Appendix A), the ECLS-K (kindergarten cohort for 1998-1999; see Appendix A), and selected states' curriculum standards.

Because of small sample sizes, the NCES longitudinal surveys program itself is able to provide only national estimates; also, entering kindergarten students are tested only periodically as samples for new cohorts are drawn. The national portrait of equity on early literacy and numeracy skills provided by ECLS-K:2011 tests could form a model for continued, standardized testing nationwide. It would be important to balance the need for age-appropriate individual assessment tools against the need for tools that could be feasibly used at a nationwide scale.

Indicator 2: Disparities in Self-Regulation and Attention Skills

As with early literacy and numeracy skills, there is at present no data source for developing measures of the two constructs under this indicator—self-regulation and attention skills for students entering kindergarten. The ECLS-K:2011 included direct assessments of kindergartner's social skills (e.g., social interaction, attentional focus, and self-control) and problem behaviors (e.g., impulsivity and externalizing problem behaviors). It also included parents' and teachers' assessments of such learning behaviors as the ability to keep belongings organized and work independently. Some states and districts have independently developed assessments of kindergartners' behavioral readiness skills. Research and development could perhaps result in streamlined assessments for use in schools nationwide, but the road to that outcome would be challenging.

DOMAIN B: K–12 LEARNING AND ENGAGEMENT

Monitoring students' achievement as they progress through K–12 is essential for signaling whether they are on track or whether interventions are needed for one or more student groups of interest. From the available research (see Chapter 4), the committee identified three indicators for which measures of constructs are needed for several grades, such as grades 4, 8, and 10, or for levels, such as elementary school, middle school, and after the first year of high school. The three indicators are engagement (Indicator 3), performance in course work (Indicator 4), and performance on tests (Indicator 5).

Indicator 3: Disparities in Engagement in Schooling

For one of the two constructs under this indicator—academic engagement—data are not readily available on the scale that is needed to develop measures as students proceed through K–12. However, there are developments that may change the picture. The National Center on Safe, Supportive, and Learning Environments, operated by the American Institutes for Research (AIR) for the U.S. Department of Education's Office of Safe and Healthy Schools, has a data collection initiative that is relevant.

In 2015 NCES piloted what is now called the ED School Climate Surveys (EDSCS) in 50 schools. Based on this work, the AIR center offers tested survey instruments and a data reporting platform to states, school districts, and schools to survey school climate as seen by 5th- to 12th-grade students, staff, and parents. The center maintains a list with links to school climate surveys conducted by states, and NCES itself plans to conduct a national-level survey. The EDSCS includes engagement as a topic, including measures of relationships among students, teachers, families, and schools, participation in school, and respect for diversity. At present, the EDSCS is provided as a resource, with an explicit promise that a jurisdiction's results will not be seen by the U.S. Department of Education unless the jurisdiction chooses to make its results public.[1]

For the other construct under Indicator 3—attendance/absenteeism, which captures the inverse of student engagement—relevant data are available. The Civil Rights Data Collection (CRDC) program in the U.S. Department of Education's Office of Civil Rights (see Appendix A) collected information on chronic absenteeism (defined as missing 15 or more days of school a year) for the 2013-2014 and 2015-2016 school years. Going forward, data on chronic absenteeism are being collected annually through ED*Facts* as part of each state's reporting requirements under the Elementary and Secondary Education Act, as amended by the Every Student Succeeds

---

[1]For information on the EDSCS, see https://safesupportivelearning.ed.gov/edscls.

Act (ESSA): the ED*Facts* definition is missing 10 percent or more of school days.[2,3]

The data in ED*Facts* could be used to develop appropriate measures of engagement (disengagement) for students categorized by gender, race and ethnicity, disability status, and English-language learner status. The location of schools (e.g., urban, suburban, town, rural) could also be used as a reporting classification, as could school level (elementary, middle, secondary, other—the data are not collected by grade). Although ED*Facts* does not collect data on absenteeism for students classified by a measure of socioeconomic status, it could be possible to use poverty estimates from the American Community Survey (ACS; see Appendix A) for school districts and school attendance areas as a proxy (see discussion under Indicator 8, below).

Indicator 4: Disparities in Performance in Coursework

Chapter 4 reviewed the literature on the role that continued academic success in school courses plays in enabling students to graduate on time from high school and be ready for college or other postsecondary pursuits. From this review, the committee concluded that measures of the following three constructs, obtained at several grades or levels, would most appropriately indicate performance in course work: success in classes, accumulating credits (being on track to graduate), and grade point average (GPA).

While the available data cannot now be used for developing measures that aggregate from student-level information, the CRDC has two directly relevant variables for groups of students. These variables are collected biannually for schools and districts and disaggregated by gender, race, disability status, and English-language learner status: number of students enrolled in and passing algebra I in middle school (separately for grades 7 and 8); and number of students enrolled in and passing algebra I in grades 9–10.[4]

Other CRDC variables that might serve as proxy measures (although they depend on school offerings—see Domain F, below) include students

---

[2]See Chang, Bauer, and Byrnes (2018) for an analysis of the CRDC data on chronic absenteeism, which puts schools into five categories, from low chronic absence (0-4.9% of students meeting the 15 or more days absent definition) to extreme chronic absence (30%+ of students meeting the definition). They find that poverty relates strongly to high rates of chronic absenteeism.

[3]The Council of the Great City Schools includes absenteeism among its *Academic Key Performance Indicators* based on a survey of its members (see Appendix B, Table B-3). Absenteeism—from the CRDC to date— is also an indicator in the NCES publication, *Status and Trends in the Education of Racial and Ethnic Groups* (see Appendix B, Table B-7).

[4]The Council of the Great City Schools includes credit for algebra I in middle school in its *Academic Key Performance Indicators* based on a survey of its members (see Appendix B, Table B-3).

enrolled in gifted and talented programs, in the International Baccalaureate (IB) Diploma Program, and in at least one Advanced Placement (AP) course and who took an AP exam.[5] The CRDC also collects data on the number of students retained in each grade, which could be used as a measure of lack of progress in course work.

At present, readily available information on course passing, accumulation of credits, and GPA is only available in national-level data sources, such as NAEP transcript studies, transcript studies conducted as part of various NCES longitudinal surveys, and parental reports in the Parent and Family Involvement in Education Survey, conducted periodically as part of the National Household Education Surveys (see Appendix A). As more states fully develop their Statewide Longitudinal Data Systems (SLDS, see Appendix A) and arrange for access to the data for statistical purposes, it should be very possible to develop the three measures of passing courses, accumulating credits, and GPA.

Indicator 5: Disparities in Performance on Tests

Given the focus over the past few decades on testing and assessment of student performance, there is a plethora of data on student test scores on reading, math, and other subjects for schools and school districts. The problem for a nationwide indicator system is that states do not all use the same tests, so that some way to make the results comparable across states is needed. The researchers behind the Stanford Education Data Archive (see Appendix A) have used NAEP test results to develop calibration factors for interpreting state test results. These factors, applied to state test scores, make the adjusted state scores more nearly comparable across states. For example, one state's test scores might need to be multiplied by a factor of 0.9 because the NAEP results indicate that that state's test gives higher scores than are justified by how students in the state perform on NAEP. Conversely, another state's test scores might need to be multiplied by a factor of 1.1 because the NAEP results indicate that that state's test gives lower scores than are merited by how the students in the state perform on NAEP. For purposes of an ongoing educational equity indicator system, there would be a need to periodically review the calibrations to take account of changes in state tests. Another issue for presentation of measures is how best to display the results: for example, in terms of the percentage of

---

[5]The Council of the Great City Schools includes participation in and passing of AP courses in its *Academic Key Performance Indicators* based on a survey of its members (see Appendix B, Table B-3). *America's Children*, produced by the Federal Interagency Forum on Child and Family Statistics, includes percentages of high school students enrolled in selected mathematics and science courses as an indicator from the CRDC (see Appendix B, Table B-5).

students scoring higher than the proficient level or by using another metric. Relatedly, there is the issue of how to capture improvement (or not) over time (see discussion of Indicator 5 in Chapter 4).

DOMAIN C: EDUCATIONAL ATTAINMENT

In Chapter 4 the committee identifies two outcomes of the K–12 education system for which it is important to measure equity among student groups of interest: on-time high school graduation (Indicator 6) and post-secondary readiness (Indicator 7).

Indicator 6: Disparities in On-Time Graduation

The current standard for measuring high school graduation rates, developed by NCES after research and consultation with stakeholders and introduced for the 2010-2011 school year, is the adjusted cohort graduation rate (ACGR). The ACGR represents the percentage of students in a state (adjusted for migration) who enter the 9th grade and earn a regular diploma in that state within 4 years. The measure is also calculated for school districts and schools. As an example, a school's ACGR for the school year 2017-2018 would be calculated as:

$$\frac{\text{Number of students earning a regular high school diploma by the end of school year 2017-2018}}{\substack{\text{Number of first-time 9th graders in the school in fall 2014} \\ \text{plus} \\ \text{Number who transferred in from other schools in school years} \\ \text{2014-2015, 2015-2016, 2016-2017, and 2017-2018} \\ \text{minus} \\ \text{Number who transferred out, emigrated, or died in school years} \\ \text{2014-2015 through 2017-2018}}}$$

ACGR rates are readily available for school districts and states from the Common Core of Data (CCD; see Appendix A) for most student groups of interest, and the rates are included in many publications (see Appendix B). Rates are also available for schools from EDFacts but not broken down for student groups of interest. Presumably, school rates for student groups of interest could be made available from the SLDS.

Indicator 7: Disparities in Postsecondary Readiness

The committee concluded (see Chapter 4) that perhaps the most useful construct to measure regarding disparities in postsecondary readiness is whether young people are actually enrolled in college, or employed, or enlisted in the military immediately following high school graduation. The ACS provides information for the nation, states, and school districts on education and employment status; however, this information cannot be tied back to the individual's high school. The most promising source for a useful measure would likely be the SLDS in states that are tracking students beyond high school.

DOMAIN D: EXTENT OF RACIAL, ETHNIC, AND ECONOMIC SEGREGATION

Indicator 8: Disparities in Students' Exposure to Racial, Ethnic, and Economic Segregation

As discussed in Chapter 5, to capture fully the aspects of school segregation that can adversely affect student outcomes and increase resource needs for schools, the committee concluded that it would be useful to develop measures for two constructs—namely, concentration of poverty in schools and racial segregation within and across schools.

For concentration of poverty in schools, a widely used measure to date has been the percentage of students eligible for free and reduced-price school lunches under the National School Lunch Program (NSLP). This measure, however, is less and less useful for this purpose for several reasons: the eligibility thresholds for reduced-price and free lunches are 185 percent and 130 percent, respectively, of the official poverty threshold; the percentage of enrolled students may vary as a function of the outreach and encouragement of each school and district to eligible families; enrollment tends to drop off with age due to stigma for older students; and more and more schools and districts are taking advantage of a provision in the NSLP program to provide free lunches to all students in schools with high percentages of eligible students in order to reduce the burden and stigma of application and verification.[6]

What would be preferable for schools is a direct measure of the percentage of poor students to use to assign schools to a few categories—say, low, medium, and high percentages of poor students, tagged as, say, "little poverty," "less concentrated poverty," and "highly concentrated poverty." Then, for multi-school districts, states, and the nation, the measure would

---

[6]See https://nces.ed.gov/blogs/nces/post/free-or-reduced-price-lunch-a-proxy-for-poverty.

be the percentages of students attending schools in each category. With knowledge of school attendance areas and how they correspond to census tracts and block groups, it would be possible to use estimates from the Small-Area Income and Poverty Estimates program (SAIPE; see Appendix A)—or estimates constructed using SAIPE methods—to categorize schools where the attendance boundaries closely overlap the geographic areas recognized in the ACS.[7] A related method, which would categorize all schools, would be to provide address information for students attending a school to the Census Bureau to keep secure and use to model the school's poverty percentage using ACS and administrative records data.

For racial segregation within and across schools, there are extensive data available in virtually every data set the committee reviewed (see Appendix A). The challenge is to develop a measure that most nearly relates to the deleterious effects of racial segregation, including determination of which racial and ethnic groups to use in the measure and which percentage values for, say, high, medium, and low racial segregation are most useful.

DOMAIN E: EQUITABLE ACCESS TO HIGH-QUALITY EARLY LEARNING PROGRAMS

Indicator 9: Disparities in Access to and Participation in High-Quality Pre-K Programs

Licensed pre-K programs include those offered by school districts, Head Start programs, and other programs licensed by their state. The CRDC provides biannual measures of whether school districts offer preschool together with the enrollment, and ages covered for most student groups of interest. Given that a sizable number of states and districts do not offer and no state requires enrollment in pre-K, a simple measure of how many students aged 3-5 are enrolled in a pre-K program offered by the district could be a barebones proxy for this indicator. The National Institute for Early Education Research (NIEER) at Rutgers University has a program to survey states about their pre-K programs, and the measures developed as part of that effort (see Appendix B, Table B-8) could suggest paths forward for this domain.

DOMAIN F: EQUITABLE ACCESS TO HIGH-QUALITY CURRICULA AND INSTRUCTION

As discussed in Chapter 5, for all students to have an equitable opportunity to succeed, school systems need to offer high-quality curricula and

---

[7]The NCES Edge program maintains information on school locations; see Appendix A.

instruction. Specifically, the committee concluded that school systems need to provide the following four things: effective teaching (Indicator 10); access to and enrollment in rigorous coursework (Indicator 11); curricular breadth (Indicator 12); and access to high-quality academic interventions and support (Indicator 13). Without these features, students will be at a disadvantage relative to other students if they wish to pursue postsecondary education and training.

Indicator 10: Disparities in Access to Effective Teaching

It is well known that teacher effectiveness matters a great deal for students' engagement with and achievement in the K–12 education system. The challenge lies in identifying constructs and measures for them that capture actual effectiveness and are feasible to obtain on a comparable basis nationwide. The committee concluded that measures of the following three constructs, obtained at several grades or levels, would most appropriately indicate effective teaching: teacher experience, teacher certification in the subjects they teach, and the racial and ethnic diversity of the teaching staff and how well teachers match their students in term of race and ethnicity.

The CRDC, biannually, and the CCD, annually, both obtain information on teacher experience and training at the school level (see Appendix A). The NCES National Teacher and Principal Survey (NTPS; see Appendix A) also obtains data on teachers' experience and training, biannually, together with their demographic characteristics such as race and ethnicity, although the sample size only permits national estimates. The SLDS in some states includes teacher as well as student characteristics, which would make it possible to construct measures of teacher diversity and teachers' racial and ethnic match with students.

Finally, states all have systems in place to measure teacher effectiveness directly, such as value-added models, student ratings in surveys, and classroom observations (see Chapter 5). It could be possible to develop a measure of teacher effectiveness using one or another of these methods, according to their use in a state. Although the resulting measures would not be uniform, they could still provide useful information.

All of these measures would need to be constructed on a school basis. Corresponding measures for multi-school districts, states, and the nation would be the percentage of students in each group of interest attending schools with effective teachers. For example, a measure might be the percentage of students attending schools that have low, moderate, or high percentages of teachers with at least a specified number of years teaching.

Indicator 11: Disparities in Access to and Enrollment in Rigorous Coursework

Opportunities to successfully enroll in and complete postsecondary education or training are very dependent on having access to required preparatory courses starting in middle school—for example, access to algebra courses in middle school or the first year of high school at the latest and access to AP or IB courses in high school. If students, particularly English-language learners, are tracked into less rigorous courses, it can be a barrier to achievement.

The CRDC obtains relevant measures biannually, including availability of and enrollment in AP courses, IB courses, and dual enrollment programs and enrollment in algebra 1 in grades 7–8 and in grades 9–10 (see Appendix A). Information on tracking is not readily available. As with Indicator 10 (above), measures of access to rigorous coursework would need to be constructed at the school level, with measures for larger areas cast in terms of the percentages of students (in each group of interest) enrolled in schools with the applicable characteristic, such as availability of AP classes. Enrollment measures for larger areas could be constructed in terms of percentages of students attending schools with no, low, medium, and high percentages of students enrolled in, say, AP courses. Alternatively, they could be constructed more directly, as the percentage of students enrolled in, say, AP courses among students attending schools that offer such courses.

Indicator 12: Disparities in Curricular Breadth

Chapter 5 discusses the value of a broad curriculum, covering much more than reading, math, and standardized test preparation, for students' educational achievement and ability to function well as adults. As the committee concluded, although it is not known which specific combination of courses is best for students' long-term outcomes, no educational system should differentially deprive students of exposure to a broad range of subjects. A measure of curricular breadth could be developed by examining state standards for subject offerings and determining the extent to which schools serving less advantaged students either do not offer some kinds of courses at all (e.g., social studies, art, a broad range of languages) or spend less time on courses other than reading and math in comparison with schools serving more advantaged students.

The CRDC biannually collects data for high schools (grades 9–12) on the number of classes in biology, chemistry, and physics combined and enrollment in those classes; it also collects data on the number of computer science classes and their enrollment. Other than those data, information on classes in social studies, art, languages, geography, and other subjects is not readily available, for elementary, middle, or high schools. It is possible that measures could be constructed from the SLDS.

Indicator 13: Disparities in Access to High-Quality Academic Supports

In addition to effective teaching, access to rigorous coursework, and curricular breadth, it is important for schools and districts to provide high-quality academic interventions and support, such as supplemental tutoring, enrichment programs or activities, additional instructional time, and personalized academic counseling, including college and career counseling. In addition, it is important for English-language learners and students with disabilities to receive the most appropriate mix of core and specialized instruction and not be isolated in instructional ghettos.

Only limited data are currently available with which to construct appropriate measures of these constructs. For example, the CRDC has information on numbers of FTE instructional aides and their aggregate salaries, which could be used to assess the additional instructional resources that are available to students (see discussion under Indicator 16, below, of how such a measure might be constructed). The CRDC also has information about access to and enrollment in various courses for student groups, including English-language learners and students with disabilities, which could help identify the extent to which these groups are receiving appropriate academic support.

DOMAIN G: EQUITABLE ACCESS TO SUPPORTIVE SCHOOLS AND CLASSROOMS

Chapter 5 discusses how supportive schools and classrooms are important for good educational outcomes, particularly for disadvantaged children. The committee identified three indicators in this domain: supportive school climates (Indicator 14); nonexclusionary discipline practices (Indicator 15); and supports for student success (Indicator 16) (other than academic supports, which are covered under Indicator 13).

Indicator 14: Disparities in School Climate

Definitions of school climate vary widely (see Chapter 5), but, in general, "climate" refers to the way that a school feels to students, the adults who work in the school, and students' families. Aspects of climate can include safety, supportiveness of staff, an academically focused culture, absence of harassment and discrimination, connectedness among students and staff, sense of fairness, and trust of adults and peers.

Although climate measures are not routinely collected by schools across the United States, several states have adopted climate measures for use in their accountability systems under the Every Student Succeeds Act, and many school districts also administer climate surveys. The National Center on Safe, Supportive, and Learning Environments in the U.S. Department of Education, has an initiative (see description under Indicator 3, above) to provide tested questions and other aids to states to administer climate surveys of middle and high school students, instructional staff, and parents or guardians. The three topic areas for which questions are available include engagement (see Indicator 3, above), safety, and the school environment. The relevant topics under safety include emotional safety, physical safety, bullying/cyberbullying, and substance abuse; the relevant topics under environment include physical environment and instructional environment. Another topic under environment is disciplinary practices, which is relevant to Indicator 15 (see below).

The CRDC currently provides comparable data for all schools and districts on some aspects of school climate. Specifically, there are data on harassment and bullying and school safety, which could be aggregated into one or more scales, based on research into which factors most strongly affect student outcomes. These data are provided by school administrators, so they represent documented instances of, for example, harassment and bullying or various kinds of violence. As such, they represent limited measures of school climate. More robust measures that capture the full spectrum of school climate to use to categorize schools as having, say, a strongly supportive climate, moderately supportive climate, or hostile climate, would require work with the states and the National Center on Safe, Supportive, and Learning Environments to develop survey measures that are as comparable as possible across jurisdictions and feasible to administer at a nationwide scale.

Indicator 15: Disparities in Nonexclusionary Discipline Practices

A school's approach to student discipline can influence students' opportunities to learn (see discussion in Chapter 5). Such exclusionary discipline policies as in- or out-of-school suspension remove students from the classroom, thereby reducing their opportunities to learn and to become engaged in their school work. As a result, these practices could negatively affect student learning and other outcomes for students who are subjected to them. It is currently not possible to measure schools' use of nonexclusionary disciplinary policies, the extent to which teachers are trained to use nonpunitive approaches, or the extent to which they effectively implement these approaches.

It is possible, however, to use suspension and expulsion rates to measure the lack of nonexclusionary methods. States are required to report those rates biannually for school districts and schools to the Office of Civil Rights as part of the CRDC, including, specifically: counts of K–12 students with and without disabilities who received one or more than one out-of-school suspension or who were expelled with educational services, without educational services, or because of zero-tolerance policies. Both sets of counts are reported by race, gender, and English-language learner status. These data could be used to classify schools into categories, such as low, moderate, and high percentages of suspended and expelled students; they could also be used to report student groups (e.g., by race and ethnicity) who are suspended or expelled at rates above, about the same, or below the average for their school, district, state, and the nation.[8]

Indicator 16: Disparities in Nonacademic Supports for Student Success

As discussed in Chapter 5, schools that serve students from poor families, students lacking in English proficiency, and students with special needs due to one or more disabilities require resources to ensure that those students have an opportunity to learn and achieve. The range of supports that schools could offer to ensure student success is almost boundless—especially in schools in which the student population has multiple needs. One category of support focuses on academics, which is the focus of Indicator 13, above. Another category relates to supporting students' socioemotional development through specific curricular programs and other means. A third type of support relates to meeting the emotional and behavioral needs of students who are exposed to violence and other stressors in their homes and neighborhoods—for example, screening (using, e.g., the Adverse Childhood Experiences tool)[9] and providing onsite counseling or appropriate referral services to students. A fourth type of support addresses students' physical health—for example, through dental or medical screenings for students who otherwise may not have access to such screenings. Of course, these supports all require resources, principally for staff.

Currently, the CRDC obtains relevant counts of non-instructional staff support for schools and districts, including number of FTE school counselors, psychologists, nurses, and social workers. The CRDC also obtains total salaries funded with state or local funds for support services staff (e.g.,

---

[8]The Council of the Great City Schools includes percent of students with out-of-school suspensions by number of days suspended in its *Academic Key Performance Indicators* based on responses to a survey of its members (see Appendix B, Table B-3). The NCES report on *Status and Trends in the Education of Race and Ethnic Groups* also includes percent of students by grade who received out-of-school suspensions from the CRDC (see Appendix B, Table B-7).

[9]See https://www.samhsa.gov/capt/practicing-effective-prevention/prevention-behavioral-health/adverse-childhood-experiences.

counselors).[10] When coupled with a measure of the percentage of students in a school who are poor or classified as English-language learners or with a disability, the above information could be used to categorize schools (and the students attending them) by the ratio of their resources to their students' needs.

Working out the technical details of an appropriate measure of non-academic supports for student success would be challenging, but one approach could proceed something like the following.[11] For a within-state indicator based on state and local (but not federal) funding, start with the statewide average per pupil costs of support staff, determine an average to allocate per nonpoor, non-English-language-learning, and nondisabled students, which would be lower than the overall average, and an average to allocate per poor, English-language-learning, and disabled students, which would be higher than the overall average. Then, determine each school's proportion of extra-needs students and non-extra-needs students, apply the appropriate per pupil dollar amount to each group, and calculate the overall average for the school. Finally, examine the school's actual per pupil costs of support staff and compare it to the needs-based ratio. Schools could then be classified as having more than adequate resources for the non-academic needs of their student body, adequate resources, or less than adequate resources.

---

[10]The financial portion of the CCD, conducted by the Census Bureau, provides data on staff expenditures that are federally funded, but only at the district level and not for individual schools.

[11]Examining the approaches used by the Education Law Center and Rutgers University for the *Is Funding Fair?* series of annual reports could also be helpful to suggest useful measures of resources relative to needs (see Appendix B).

Appendix D

Agendas for Public Sessions of the Committee

Committee on Identifying Indicators of Education Equity
Committee on National Statistics (CNSTAT)
Commission Behavioral and Social Sciences and Education (DBASSE)

Meeting 1: April 20-21, 2017
National Academies of Sciences, Engineering, and Medicine
Keck Center
500 Fifth Street, NW
Washington, DC

Agenda

9:30 am-3:15 pm   OPEN SESSION

9:30 am   Welcome, Introductions

         Mary Ellen O'Connell, *Executive Director*, DBASSE
         Constance Citro, *Director*, CNSTAT

9:45   Sponsors' Goals and Priorities
        Discussion Leader: Laura Hamilton

        Each sponsor should talk about the goals they have for the project: What expectations do they have for the project? What impacts would they like to see? How can we start build-

ing momentum and political will for the adoption of equity indicators?

- **Vivian Tseng,** William T. Grant Foundation
- **Kim Robinson,** W.K. Kellogg Foundation
- **Chris Chapman,** Institute of Education Sciences (IES), U.S. Department of Education
- Spencer Foundation
- Atlantic Philanthropies
- Ford Foundation

10:30   Take-away messages from the October 5, 2015, planning meeting
             Discussion Leader: Michael MacKenzie

- **Richard Murnane,** Harvard (by telephone)
- **Natalie Nielsen,** *Staff,* Board on Testing and Assessment, DBASSE

11:00   **Break**

11:15 am-   **Moderated Conversations with Researchers and Stakeholders**
3:00 pm

11:15   **Panel 1: Considerations when Adopting Indicators**
             Discussion Leader: Jim Kemple

Panelists will discuss the processes and criteria that are used to ensure that indicators represent valid, reliable, and useful measures of the status of the education system. Panelists will draw from two concrete examples that are illustrative of ones that have moved from research use to policy use—graduation rates and achievement test results.

- **Elaine Allensworth,** Consortium of Chicago School Research (committee member): *High School Graduation Rates*
- **Sean Reardon,** Stanford (committee member): *State Achievement Test Results*

11:45   **Panel 2: Government Affiliated Data Collections**
             Discussion Leader: Sharon Lewis

Panelists from two levels of government—federal and district— will discuss uses of civil rights/equity data collected by gov-

ernment agencies. Ms. Lhamon will discuss the ways the US Commission on Civil Rights uses the data. Mr. Carvalho will talk about the data his district is required to collect for different levels of government and the extent to which it is useful for informing policy and practice at the local level

- **Catherine Lhamon**, *Chair,* U.S. Commission on Civil Rights
- **Alberto Carvalho**, *Superintendent,* Miami-Dade County Schools; *Committee Member*

| 12:25-1:15 pm | Lunch |

1:15   **Panel 3: Use of Indicators for Policy Purposes (50 min.)**
Discussion Leaders: Meredith Phillips, Karolyn Tyson

This panel includes individuals from several organizations that advocate for improvements in education for all students. Panelists will address the following questions: (1) Of the equity indicators that the committee might focus on, what types would be most useful to the aims/mission of your organization? (2) How would you use them in your work?

- **Natasha Ushomirsky**, Education Trust
- **Nat Mulkas**, American Enterprise Institute
- **Amber Northern**, Fordham Institute
- **Stephanie Wood-Garnett**, Alliance for Excellence in Education

2:05   **Panel 4: Disseminating and Interpreting Indicators (50 min.)**
Discussion Leaders: Stella Flores, Morgan Polikoff
[*10 mins per panelist, 10 min for discussion*]

The panel includes individuals from organizations that disseminate and interpret information related to education equity. Panelists will share samples of reports published by their organization and address the following questions: (1) What equity indicators do you currently report? (2) What feedback do they receive from users of your reports?

- **David Murphey**, Child Trends
- **Betsy Brand**, American Youth Policy Forum
- **Ilene Berman**, Annie E Casey Foundation

• **Jennifer Park,** Federal Interagency Forum on Child and Family Statistics: National Indicators of Well Being, U.S. Office of Management and Budget

3:00   **Summing Up**
          Christopher Edley

3:15   **Adjourn**

Committee on Identifying Indicators of Education Equity
Committee on National Statistics (CNSTAT)

Meeting 2: October 2-3, 2017
National Academies of Sciences, Engineering, and Medicine
Keck Center
Room 201
500 Fifth Street, NW
Washington, DC

Agenda

9:30 am-2:15 pm   OPEN SESSION
                     MODERATED PANEL DISCUSSIONS

9:30 am   Tab B Welcome, Introductions, Review of the agenda
             Christopher Edley, *Chair*

             Panel Discussions
             Panel discussions will focus on research related to the follow-
             ing valued education outcomes:

             1. *Kindergarten readiness*
             2. *Strong academic growth and achievement in English lan-*
                guage arts and math in grades K–12
             3. *Engagement in schooling (i.e., attendance, course enrollment)*
             4. *On-time high school graduation*
             5. *Graduating college-ready (in terms of coursework and GPA,*
                not just ACT/SAT scores)
             6. *Postsecondary enrollment in higher education and training*

Each panelist has been assigned to one or two of the six outcomes and has agreed to prepare a paper that reviews the literature on equity issues related to the outcome.

In the paper, panelists will also examine literature on the predictors/correlates of those outcomes and on equity issues related to these predictors and correlates.

Panelists will consider predictors and correlates from 4 broad categories: (a) family and home environment; (b) child and adolescent social-emotional and academic development; (c) in-school structures, supports, and resources; and (d) community and neighborhood environment.

The panel discussions are designed to lay the groundwork for the papers, and the information will be used to refine the goals for each paper.

There will be 3 parts to the panel discussions. Part 1 will focus on each outcome separately. Parts 2 and 3 will be cross-outcome discussions.

Panelists

Pre-K Readiness
Katherine Magnuson, University of Wisconsin (by telephone)

K–12: Strong Achievement and Academic Progress and Engagement in Schooling
These two outcomes were combined for the purpose of the literature reviews because they draw from similar research bases. The categories of predictors have been split across 3 panelists as detailed below.

> **Douglas Ready, Columbia Teachers College**
> *Will address predictors related to in-school supports, such as school curricula, resources, teachers, support systems*

> **Lori Taylor Texas A&M**
> *Will address predictors related to school finance, economics of education*

On-Time High School Graduation
Russell Rumberger and Jay Plasman, University of California,
Santa Barbara

Transition to Postsecondary Education & Training
Lashawn Richburg-Hayes, Insight Policy Research

9:45-11:00 Part 1: Describe Outcomes and Predictors

Discussion Leaders: Laura Hamilton, Sara McLanahan

*Each panelist will give a brief, 10-minute overview for the
assigned outcome and address the questions listed below. This
may could include a few power point slides, handouts, and/or
citations for important studies.*

1. *Describe Outcomes: How is each outcome defined and measured? What are the components/subcomponents of each
outcome?*

2. *Describe Predictors/Correlates/Opportunities: What are the
most powerful predictors/correlates of performance on the
outcome? How are the predictors/correlates defined and
measured?*

11:00 am–   Part 2: Cross-Outcomes Discussion of Equity
12:00 pm

Discussion Leaders: Sean Reardon, Sharon Lewis

*Two committee members will lead panelists in a discussion
of similarities and differences in predictors/correlates across
outcomes. The focus will be on the factors that affect disadvantaged students in ways that result in disparities in outcomes.*

3. *How do predictors/correlates relate to equity? Which
predictors/correlates are most associated with disparities
(e.g., by race, SES, immigration status, EL status, other) on
each outcome measure?*

12:00–   Lunch in Meeting Room
12:45

12:45-2:15 Part 3: Cross-Outcomes Discussion of Indicators and Policy
Interventions (90 mins.)

Moderators: Karolyn Tyson, Morgan Polikoff

*Two committee members will lead panelists in a cross-out-
comes discussion of research on leverage points for policy
interventions and on potential indicators that could facilitate
such interventions.*

*4. Which predictors/correlates are sensitive to policy interven-
tion? Which predictors/correlates are the most "malleable,"
and likely to change as a result of targeted policy interventions intended -to increase equity in the outcomes?*

*5. Which factors are candidates for indicators? What compo-
nents of the outcomes and predictors/correlates "rise to the
top" as potential indicators? What do you see as challenges
or limitations associated with those potential indicators?*

2:15   Adjourn

Appendix E

Biographical Sketches of Committee Members and Staff

Christopher Edley, Jr. (*Chair*) is the honorable William H. Orrick, Jr., distinguished professor and faculty director at the Chief Justice Earl Warren Institute on Law and Social Policy at the University of California Berkeley School of Law. Previously, he was dean of the Berkeley School of Law and a professor at Harvard Law School. His academic work is broadly in administrative law, civil rights, education policy, and domestic public policy. In addition to his academic work, he served in White House policy and budget positions in the Carter and Clinton administrations. He also held senior positions in five presidential campaigns, including as senior policy adviser for Barack Obama and on Obama's Transition Board. More recently, he cochaired the congressionally chartered National Commission on Education Equity and Excellence, and he chairs the follow-on effort, For Each & Every Child. He is a member of the American Academy of Arts & Sciences, the National Academy of Public Administration, the Council of Foreign Relations, and the Gates Foundation's National Programs Advisory Panel He has a B.A. in mathematics from Swarthmore College, an M.A. from Harvard University John F. Kennedy School of Government, and a J.D. from Harvard Law School.

Elaine Allensworth is the Lewis-Sebring director of the Consortium on Chicago School Research, where she conducts studies on what matters for student success and school improvement. Her research on early indicators of high school graduation has been used to create student monitoring systems for Chicago and other districts across the country. In addition to studying educational attainment, she conducts research in the areas of school leadership and school organization. She has received a number of awards from the American Educational Research Association (AERA) for outstanding publications, including the Palmer O. Johnson award for an outstanding article in an AERA journal, Division H awards for outstanding instructional research and planning research, and a policy and management research award She has an M.A. in sociology and urban studies and a Ph.D. in sociology, both from Michigan State University.

Alberto Carvalho is superintendent of Miami-Dade County Public Schools, the nation's fourth largest school system, with more than 350 schools serving 400,000 students. Previously, he served the school system in several capacities, including as chief communications officer, administrative director, and as both associate and assistant superintendent. Under his leadership, the district has won the College Board Advanced Placement Equity and Excellence District of the Year award, the Cambridge District of the Year award, and the Broad Prize for Urban Education. He is the recipient of the Florida Superintendent of the Year award, the Urban Superintendent of the Year award, the National Superintendent of the Year award of the American Association of School Administrators, and the Superintendent of the Year award of the National Association of Bilingual Education. He is also winner of the Harold W. McGraw Prize in Education and has received honors from both Mexico and Portugal. He currently serves on the National Assessment Governing Board. He holds a bachelor's degree in biology/biomedical sciences from Barry University and a master's degree in educational leadership from Nova Southeastern University.

Constance F. Citro is a senior scholar in the Committee on National Statistics (CNSTAT). She previously served as CNSTAT director and senior study director. She began her career with CNSTAT in 1984 as study director for the panel that produced *The Bicentennial Census: New Directions for Methodology in 1990*. Prior to joining CNSTAT, she held positions as vice president of Mathematica Policy Research, Inc., and Data Use and Access Laboratories, Inc. She was an American Statistical Association/National Science Foundation/Census research fellow in 1985-1986, and is a fellow of the American Statistical Association and an elected member of the International Statistical Institute. For CNSTAT, she directed evaluations of the 2000 census, the Survey of Income and Program Participation, microsimulation models for social welfare programs, and the National Science Foundation science and engineering personnel data system, in addition to studies on institutional review boards and social science research, estimates of poverty for small geographic areas, data and methods for retirement income modeling, and a new approach for measuring poverty. She co-edited the 2nd–6th editions of *Principles and Practices for a Federal Statistical*

Agency. She received her B.A. in political science from the University of Rochester, and M.A. and Ph.D. in political science from Yale University.

Stella Flores is an associate professor of higher education at the Steinhardt School of Culture, Education, and Human Development at New York University. In her research she uses large-scale databases and quantitative methods to investigate the effects of state and federal policies on college access and completion rates for low-income and underrepresented populations. That research covers minority-serving institutions, immigrant students, English-language learners, the role of alternative admissions plans and financial aid programs in college admissions in the United States and abroad, demographic changes in U.S. education, and Latino students and community colleges. Previously, she served as an associate professor at Vanderbilt University and held positions at the U.S. Government Accountability Office and the U.S. Economic Development Administration. She has a B.A. from Rice University, a master's degree in public affairs from the University of Texas at Austin, and an Ed.M. and an Ed.D. in administration, planning, and social policy from Harvard University.

Nancy Gonzales is associate dean of faculty at Arizona State University College of Liberal Arts and Sciences. Her primary research interests focus on cultural and contextual influences on adolescent mental health, and her areas of research include culture and ethnic issues in prevention research; prevention of Mexican-American school dropout and mental health problems; acculturation and enculturation of Mexican-American children and families; and contextual influences on adolescent development. Her work includes research on the role of neighborhood disadvantage and acculturation on children's mental health and on how these influences are mediated or moderated by family processes in Mexican-American and African-American families. She also is involved in the development and evaluation of culturally sensitive interventions for Mexican-American and African-American families. She has a Ph.D. from the University of Washington.

Laura Hamilton is a senior behavioral scientist and distinguished chair in learning and assessment at the RAND Corporation, where she directs the RAND Center for Social and Emotional Learning Research and codirects the American Educator Panels. She also serves as a faculty member at the Pardee RAND Graduate School. Previously, she served as an adjunct faculty member in the University of Pittsburgh's Learning Sciences and Policy program. Her research addresses topics related to social and emotional learning, educational assessment, accountability, the implementation of curriculum and instructional reforms, and education technology. Recent projects include a study of a social and emotional learning intervention for elementary schools and afterschool programs, the development of a database of measures of interpersonal and intrapersonal competencies, and an evaluation of personalized learning interventions. She has also served on several state and national panels on topics related to assessment, accountability, and educator evaluation. She has an M.S. in statistics and a Ph.D. in educational psychology from Stanford University.

James Kemple is the executive director of the Research Alliance for New York City Schools and research professor at the Steinhardt School of Culture, Education, and Human Development at New York University. For the Research Alliance, he serves as the principal investigator on a range of studies, including those examining the efficacy of on-track indicators for different grade levels; performance trends in New York City (NYC) high schools; and the effects of school closure. His work focuses on examining high school reform efforts, assessing performance trends in the city's educational landscape, and designing rigorous impact evaluations. He collaborates with the NYC Department of Education, private foundations, and other stakeholders to identify research priorities and develop new lines of inquiry. Earlier in his career, he was a high school mathematics teacher, and he also managed the Higher Achievement Program, which serves disadvantaged youth in Washington, D.C. He has a B.A. in mathematics from the College of the Holy Cross and an Ed.M. and an Ed.D. from the Harvard University Graduate School of Education.

Judith Koenig (*Study Director*) is on the staff of the Committee on National Statistics of the National Academies of Science, Engineering, and Medicine, where she directs measurement-related studies designed to inform education policy. Her work has included studies on the National Assessment of Educational Progress; teacher licensure and advanced-level certification; inclusion of special-needs students and English-language learners in assessment programs; setting standards for the National Assessment of Adult Literacy; assessing 21st-century skills; and using value-added methods for evaluating schools and teachers. Previously, she worked at the Association of American Medical Colleges and as a special education teacher and diagnostician. She has a B.A. in elementary and special education from Michigan State University, an M.A. in psychology from George Mason University, and a Ph.D. in educational measurement, statistics, and evaluation from the University of Maryland.

Sharon Lewis recently retired from the position of director of research for the Council of the Great City Schools in Washington, D.C. In that position she had directed the council's research program, which contributes to the organization's efforts to improve teaching and learning in the nation's urban schools and to help develop education policy. She previously worked as a national education consultant and as assistant superintendent of research, development, and coordination with the Detroit Public Schools. She has an M.A. in educational research from Wayne State University.

Michael J. MacKenzie is the Canada research chair on child well-being and professor of social work, psychiatry, and pediatrics at McGill University. His research focuses on the accumulation of stress and adversity in early childhood and the impact on caregiver perceptions and subsequent parenting behavior, including the roots of maltreatment and the pathways of children into and through the child welfare system. He served as the principal investigator on a UNICEF-funded project in the Hashemite Kingdom of Jordan that represented one of the first implementations of foster care and juvenile diversion as community-based alternatives to institutionalization in the region. His honors include a William T. Grant Foundation faculty scholar award to support a project examining the biological and social underpinnings of the placement trajectories and well-being of children in the foster care system and an excellence in research award from the Society for Social Work and Research. He has a B.Sc. and M.Sc. from the University of Western Ontario and an M.S.W., M.A., and Ph.D. from the University of Michigan.

C. Kent McGuire is program director for education at the William and Flora Hewlett Foundation. Previously, his positions included president and CEO of the Southern Education Foundation, dean of the College of Education and professor in the Department of Educational Leadership and Policy Studies at Temple University, senior vice president at Manpower Demonstration Research Corporation, and education program officer at the Pew Memorial Trust and at the Eli Lilly Endowment. He also served in the Clinton administration as assistant secretary of education, focusing on research and development. His current research interests focus on education administration and policy and organizational change. He has also been involved in a number of evaluation research initiatives on comprehensive school reform, education finance, and school improvement. He has a master's degree in education administration and policy from Teachers College at Columbia University and a Ph.D. in public administration from the University of Colorado at Boulder.

Sara McLanahan is the William S. Tod professor of sociology and public affairs at Princeton University. She is the founding director of the Bendheim-Thoman Center for Research on Child Wellbeing and the interim director of the center's Education Research Section. She is a principal investigator of the Fragile Families and Child Wellbeing Study and editor-in-chief of *The Future of Children*, a journal dedicated to providing research and analysis to promote effective policies and programs for children. She currently serves on the Board of Trustees for the Russell Sage Foundation, and she is a past president of the Population Association of America and a past member of the boards of the American Sociological Association and the Population Association of America. She is an elected member of the National Academy of Sciences, the American Academy of Political Science, and the American Philosophical Society. She has a Ph.D. in sociology from the University of Texas at Austin, and she is the recipient of an honorary degree from Northwestern University.

Natalie Nielsen is an independent research and evaluation consultant whose work focuses on improving opportunities and outcomes for young people. Before becoming an independent consultant, she spent 5 years at the National Academies of Sciences, Engineering, and Medicine, first as a senior program officer for the Board on Science Education and later as the acting director of the Board on Testing and Assessment. She also served as the director of research at the Business-Higher Education Forum and as a senior researcher at SRI International. Nielsen holds a Ph.D. in education from George Mason University, an M.S. in geological sciences from San Diego State University, and a B.S. in geology from the University of California, Davis.

Meredith Phillips is associate professor of public policy and sociology at the Luskin School of Public Affairs at the University of California at Los Angeles. Her work focuses on the causes and consequences of educational inequality, particularly on the causes of ethnic and socioeconomic disparities in educational success and how to reduce those disparities. Her current research projects include a random-assignment evaluation of the efficacy of two low-cost college access interventions and an ethnographic longitudinal study of adolescent culture, families, schools, and academic achievement. With colleagues, she recently developed school and classroom environment surveys for the Los Angeles Unified School District. She cofounded EdBoost, a charitable, educational nonprofit whose mission is to reduce educational inequality by making high-quality supplemental educational services accessible to children from all family backgrounds. Phillips also cofounded and serves as research advisor to the Los Angeles Education Research Institute. She has an A.B. from Brown University and a Ph.D. from Northwestern University.

Morgan Polikoff is an associate professor of education at the Rossier School of Education at the University of Southern California. His areas of research include K–12 education policy; Common Core standards; assessment policy; alignment among instruction, standards, and assessments; and the measurement of classroom instruction. He uses quantitative methods to study the design, implementation, and effects of standards, assessment, and accountability policies. Recent work has investigated teachers' instructional responses to content standards and critiqued the design of school and teacher accountability systems. His ongoing work focuses on the implementation of Common Core standards and the influence of curriculum materials and assessments on implementation. He has a bachelor's degree in mathematics and secondary education from the University of Illinois at Urbana-Champaign and a Ph.D. with a focus on education policy from the University of Pennsylvania's Graduate School of Education.

Sean F. Reardon is professor of poverty and inequality in education and a professor of sociology at Stanford University. His research focuses on the causes, patterns, trends, and consequences of social and educational inequality, the effects of educational policy on educational and social inequality, and applied statistical methods for educational research. He also develops methods of measuring social and educational inequality, including the measurement of segregation and achievement gaps, as well as methods of causal inference in educational and social science research. He teaches graduate courses in applied statistical methods, with a particular emphasis on the application of experimental and quasi-experimental methods to the investigation of issues of educational policy and practice. He is a member of the National Academy of Education and the American Academy of Arts and Sciences. He has been a recipient of a William T. Grant Foundation Scholar Award and a Carnegie Scholar Award. He has a Ph.D. in education from Harvard University.

Karolyn Tyson is an associate professor in the Department of Sociology at the University of North Carolina at Chapel Hill. She specializes in qualitative research focused on issues related to schooling and inequality. She is particularly interested in understanding the complex interactions between schooling processes and the achievement outcomes of black students. Some of her current and recent work includes a multi-method, multi-site study examining issues centered on the law, rights consciousness, and legal mobilization in American secondary schools; an examination of how and why black students have come to equate school success with whiteness; and a study tracing the history of racialized tracking in a suburban school district and its consequences for the district's black students. She has a B.A. from Spelman College and a Ph.D. in sociology from the University of California at Berkeley.

COMMITTEE ON NATIONAL STATISTICS

The Committee on National Statistics was established in 1972 at the National Academies of Sciences, Engineering, and Medicine to improve the statistical methods and information on which public policy decisions are based. The committee carries out studies, workshops, and other activities to foster better measures and fuller understanding of the economy, the environment, public health, crime, education, immigration, poverty, welfare, and other public policy issues. It also evaluates ongoing statistical programs and tracks the statistical policy and coordinating activities of the federal government, serving a unique role at the intersection of statistics and public policy. The committee's work is supported by a consortium of federal agencies through a National Science Foundation grant, a National Agricultural Statistics Service cooperative agreement, and several individual contracts.